Communications
in Computer and Information Science 1480

More information about this series at http://www.springer.com/series/7899

Yongtian Wang · Weitao Song (Eds.)

Image and Graphics Technologies and Applications

16th Chinese Conference
on Image and Graphics Technologies, IGTA 2021
Beijing, China, June 6–7, 2021
Revised Selected Papers

Springer

Editors
Yongtian Wang
Beijing Institute of Technology
Beijing, China

Weitao Song
Beijing Institute of Technology
Beijing, China

ISSN 1865-0929   ISSN 1865-0937 (electronic)
Communications in Computer and Information Science
ISBN 978-981-16-7188-3   ISBN 978-981-16-7189-0 (eBook)
https://doi.org/10.1007/978-981-16-7189-0

This Springer imprint is published by the registered company Springer Nature Singapore Pte Ltd.
The registered company address is: 152 Beach Road, #21-01/04 Gateway East, Singapore 189721, Singapore

Preface

It was a great honor for us, the Beijing Society of Image and Graphics, to organize the 16th Conference of Image and Graphics Technology and Application (IGTA 2021) held at the Beijing Institute of Technology on June 6, 2021, which was a great success and attracted extensive attention from the academic community. For the last two years, COVID-19 has changed our way of life and brought far-reaching effects to the whole world. However, the rampant pandemic cannot stop our enthusiasm for academic exchange. We therefore gathered at IGTA 2021 to share and exchange opinions on the latest progress achieved in the past year.

IGTA is a professional meeting and an important forum for image processing, computer graphics, and related topics including, but not limited to, image analysis and understanding, computer vision and pattern recognition, data mining, virtual reality and augmented reality, and image technology applications.

This year, we received 86 technical submissions from authors in different countries and regions. Despite of the influence of the COVID-19 pandemic, the paper selection remained highly competitive this year. Each manuscript was reviewed by at least two reviewers, and some of them were reviewed by three reviewers. After careful evaluation, 21 manuscripts were selected for oral and poster presentations.

The keynote, oral, and poster presentations of IGTA 2021 reflected the latest progress in the field of image and graphics, and the papers are included in this volume of proceedings, which we believe will provide valuable references for scientists and engineers in related fields.

As the conference general chairs and program chairs, we are very grateful to the Program Committee and Organizing Committee members for their support, for the time spent to review and discuss the submitted papers, and for doing so in a timely and professional manner.

We would like to express our sincere gratitude to the authors of the excellent papers accepted for submitting and presenting their works at the conference, and to all the conference attendees for making IGTA an excellent forum on image and graphics, facilitating the exchange of ideas, fostering new collaborations, and shaping the future of this exciting research field. We hope the readers will find in these pages interesting material and fruitful ideas for their future work.

When everybody adds fuel, the flames rise high. We believe that, with our joint efforts, we will be able to conquer the pandemic, and we look forward to seeing you face to face at IGTA 2022 next year!

June 2021

Yongtian Wang
Weitao Song

Organization

General Conference Chairs

Wang Yongtian	Beijing Institute of Technology, China
Song Weitao	Beijing Institute of Technology, China

Executive and Coordination Committee

Wang Shengjin	Tsinghua University, China
Huang Qingming	University of Chinese Academy of Sciences, China
Zhao Yao	Beijing Jiaotong University, China
Chen Chaowu	The First Research Institute of the Ministry of Public Security of P.R.C, China
Liu Chenglin	Institute of Automation, Chinese Academy of Sciences, China
Zhou Mingquan	Beijing Normal University, China
Jiang Zhiguo	Beihang University, China

Program Committee Chairs

Peng Yuxin	Peking University, China
He Ran	Institute of Automation Chinese Academy of Sciences, China
Ji Xiangyang	Tsinghua University, China

Organizing Chairs

Liu Yue	Beijing Institute of Technology, China
Hai Huang	Beijing University of Post and Telecommunications, China
Yuan Xiaoru	Peking University, China
Li Xueming	Beijing University of Post and Telecommunications, China

Research Committee Chairs

Liang Xiaohui	Beihang University, China
Dong Jing	Institute of Automation Chinese Academy of Sciences, China
Yang Jian	Beijing Institute of Technology, China

Publicity and Exhibition Committee Chairs

Yang Lei   Communication University of China, China
Zhang Fengjun   Software Institute of the Chinese Academy of Sciences,
                          China
Zhao Song   Shenlan College, China

Program Committee Members

Henry Been-Lirn Duh   La Trobe University, Australia
Takafumi Taketomi   NAIST, Japan
Jeremy M. Wolfe   Harvard Medical School, USA
Huang Yiping   Taiwan University, China
Youngho Lee   Mokpo National University, South Korea
Ma Zhanyu   Beijing University of Post and Telecommunications,
                          China
Nobuchika Sakata   Osaka University, Japan
Seokhee Jeon   Kyung Hee University, Korea
Wang Yongtian   Beijing Institute of Technology, China
Shi Zhenwei   Beihang University, China
Xie Fengying   Beihang University, China
Wang Shengjin   Tsinghua University, China
Zhao Yao   Beijing Jiaotong University, China
Huang Qingming   University of Chinese Academy of Sciences, China
Chen Chaowu   The First Research Institute of the Ministry of Public
                          Security of P.R.C, China
Zhou Mingquan   Beijing Normal University, China
Jiang Zhiguo   Beijing University of Aeronautics and Astronautics,
                          China
Yuan Xiaoru   Peking University, China
He Ran   Institute of Automation, Chinese Academy of Sciences,
                          China
Yang Jian   Beijing Institute of Technology, China
Ji Xiangyang   Tsinghua University, China
Liu Yue   Beijing Institute of Technology, China
Ma Huimin   Tsinghua University, China
Zhou Kun   Zhejiang University, China
Tong Xin   Microsoft Research Asia, China
Wang Liang   Institute of Automation, Chinese Academy of Sciences,
                          China
Zhang Yanning   Northwestern Polytechnical University, China
Zhao Huijie   Beijing University of Aeronautics and Astronautics,
                          China
Zhao Danpei   Beijing University of Aeronautics and Astronautics,
                          China

Yang Cheng	Communication University of China, China
Yan Jun	Journal of Image and Graphics, China
Xia Shihong	Institute of Computing Technology, Chinese Academy of Sciences, China
Cao Weiqun	Beijing Forestry University, China
Di Kaichang	Institute of Remote Sensing and Digital Earth, Chinese Academy of Sciences, China
Yin Xucheng	University of Science and Technology Beijing, China
Gan Fuping	Ministry of Land and Resources of the People's Republic of China, China
Lv Xueqiang	Beijing Information Science & Technology University, China
Liu Wenping	Beijing Forestry University, China
Liu Jianbo	Communication University of China, China
Lin Hua	Tsinghua University, China
Lin Xiaozhu	Beijing Institute of Petrochemical Technology, China
Li Hua	Institute of Computing Technology, Chinese Academy of Sciences, China
Jiang Yan	Beijing Institute of Petrochemical Technology, China
Dong Jing	Institute of Automation, Chinese Academy of Sciences, China
Sun Yankui	Tsinghua University, China
Zhuo Li	Beijing University of Technology, China
Li Qingyuan	Chinese Academy of Surveying and Mapping, China
Yuan Jiazheng	Beijing Union University, China
Zhang Aiwu	Capital Normal University, China
Cheng Mingzhi	Beijing Institute of Technology, China
Wang Yahui	Beijing University of Graphic Communication, China
Yao Guoqiang	Beijing Film Academy, China
Ma Siwei	Peking University, China
Liu Liang	Beijing University of Posts and Telecommunications, China
Jiang Yan	Beijing Institute of Fashion Technology, China

Contents

**Other Research Works and Surveys Related to the Applications
of Image and Graphics Technology**

Image Processing and Enhancement Techniques

Residual Multi-resolution Network for Hyperspectral Image Denoising

Shiyong Xiu, Feng Gao[✉], and Yong Chen

School of Information Science and Engineering, Ocean University of China,
Qingdao, China
gaofeng@ouc.edu.cn

Abstract. Hyperspectral image (HSI) denoising is an important tool to improve the quality of HSIs for subsequent tasks. In this paper, we propose a novel method based on Residual Multiresolution Network (RMR-Net) for HSI denoising, which exploits multiscale information better from multiresolution versions of HSIs produced by pixelshuffle operation. The convolutional neural network (CNN) is used for extracting the spatial information among different resolution HSI, respectively. Enhanced representation will be obtained by fusing these multiresolution features. Wide receptive fields are provided by dilated convolution. Spectral information is also considered in the proposed network. To ease the flow of low-frequency information, we use a residual structure in our method. In addition, the experiment results on the simulated dataset demonstrate the superiority of our RMRNet.

Keywords: Hyperspectral image (HSI) denoising · Convolutional neural network · Multi-resolution · Adjacent spectral correlation

1 Introduction

Hyperspectral images (HSIs), which are made up of both spatial and spectral information of real world scenes, have already been used for many remote sensing tasks, such as classification [1] and change detection [2]. High quality HSIs are essential to improve the performance in the above-mentioned applications. Nevertheless, HSIs commonly suffer from noise during the acquisition process because of the limited light, sensor internalmalfunction, photon effects, and atmospheric interference. Therefore, it is a critical process to reduce the noise of the HSIs before the image analysis and interpretation.

Many image denoising methods have been conducted over the last decades, such as non-local means (NLM) [3] and block-matching and 3-D filtering (BM3D) [4]. NLM and BM3D search 2-D image fragments similar to the reference patch within a specific range of the image for denoising. These 2-D image denoising methods could also be employed to HSI denoising band by band. However, different from the RGB image, the HSIs could deliver spectral information of the

© Springer Nature Singapore Pte Ltd. 2021
Y. Wang and W. Song (Eds.): IGTA 2021, CCIS 1480, pp. 3–9, 2021.
https://doi.org/10.1007/978-981-16-7189-0_1

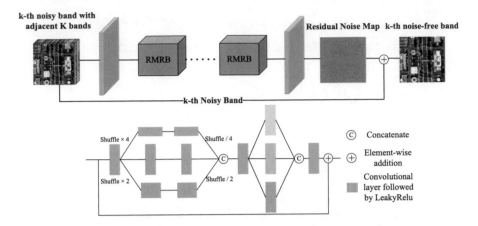

Fig. 1. Framework of the proposed RMRNet for HSI denoising. The main component of the RMRNet is the residual multi-resolution block (RMRB) which is used for capturing the features among different resolution and the spatial-spectral information in HSIs. It also allows information fusion across parallel streams via shuffling and unshuffling operations in order to consolidate the high-resolution features with the help of low-resolution information. Different green colors of the convolution layers denote different dilations. (Color figure online)

real world scenes. Therefore, bandwise 2-D denoising methods will cause spectral distortion if they are directly employed on HSIs, since they do not consider the spectral correlations among different bands. To solve this problem, some 3-D denoising methods were proposed for HSIs denoising. Maggioni et al. [5] proposed BM4D which is an extended version of the BM3D filter to 3-D data. Some low-rank matrix recovery approaches [6,7] are proposed by performing both the spatial low-rank approximation and spectral dimensionality reduction. Although these methods produce satisfying performance by effectively utilizing the underlying characteristics of HSIs, most of them formulate the HSI denoising as a complex optimization problem to be solved iteratively. Therefore, they are commonly time-consuming and unintelligent.

Recently, deep learning-based methods have achieved great success in computer vision and natural language processing tasks. They commonly use an end-to-end strategy to solve complex tasks without any human handcrafted prior, and provide an effective yet time-saving tool to tackle tricky problems. Several deep learning-based image denoising methods have been proposed, such as residual dense network [8] and encoder-decoder network [9]. Convolutional neural networks (CNNs) are employed to exploit feature representations, which have achieved significant success on image denoising. Nevertheless, these methods lack knowledge about spatial-spectral correlation and global correlation along the spectrum in HSIs.

In this paper, we propose a Residual Multi-Resolution Network (RMRNet) for HSI denoising. Both spatial and adjacent correlated spectral information is taken into account. The proposed method is capable of reducing the noise from different HSI sensors, and it can simultaneously preserve the details and structural information. To better exploit the correlation between different scales, multiresolution versions of the HSIs are considered in the proposed network, and then multiresolution feature representations are integrated to produce noise-free HSIs.

2   Methodology

2.1   Network Architecture

The framework of the proposed method is described in Fig. 1. The network is based on a recursive residual structure. The residual multiresolution block is the basic component in RMRNet. To exploit more spatial-spectral information, simulated kth noisy band and its adjacent K bands are taken as the inputs into the proposed network to produce the residual map. The input data of the size $(K + 1) \times W \times H$ represents the current noisy band with the current spatial-spectral cube with K adjacent bands. In addition, both the initial and final output features are composed of one convolutional layer.

We optimize the proposed network using the $L1$ loss:

$$\mathcal{L}(w) = \frac{1}{N} \sum_{i}^{N} \|\text{RMRNet}(x_i) - y_i\|_1 \tag{1}$$

where N is the batch size, x is the noisy input and y is the ground truth, w denotes all trainable network parameters.

2.2   Residual Multi-Resolution Block

The Residual Multi-Resolution Block (RMRB) is the critical part of the proposed network. It is comprised of two components: 1) Parallel convolution streams for multiresolution HSIs generated by the shuffling operation to extract rich multiscale and spatially features; 2) three branch convolutions to provide a wide receptive field through kernel dilation.

The main idea of the multiresolution branch is taking advantage of the low-resolution information to guide the HSI denoising process at high-resolution scale. Many tasks like semantic segmentation obtain multiscale features by applying down-sampling operation or convolution layers with different strides to the original input. However, more loss of fine details and higher calculation are caused by these methods. Compared with other strategies, shuffling is able to reduce spatial resolution but keeps all the information of inputs at the same time. Therefore, we use shuffling instead of down-sampling or convolution operations to generate multiresolution versions as

$$\mathcal{P}(I_{H,W,C}) = I'_{\lfloor H/r \rfloor, \lfloor W/r \rfloor, C \cdot r \cdot mod(H,r) + C \cdot r \cdot mod(W,r)} \tag{2}$$

where, I is the input feature of size $H \times W \times C$, shuffling operation $\mathcal{P}$ produce I' with spatial dimension $H/r \times W/r \times r^2 C$, r is the ratio of the operation. Since the low-resolution of I' is r times smaller than that of the original input, the convolution layer in low-resolution is able to capture large-scale contextual information. Having the lower resolution information, we propagate it to the higher resolution to help the feature extraction. Finally, the multiresolution information will be gathered at the original resolution to generate the enhanced representation of the HSI. As shown in Fig. 1, two Conv+LeakyReLU layers are used as feature extraction within each branch. In addition, one convolution is employed as feature transformation at the beginning of the RMRB, features from different resolution branches are fused by concatenation.

To capture features from wide receptive fields, we use kernel dilation in each RMRB for three branch convolutions. Compared with general convolution, dilated convolution can capture large-scale correlation without increasing the computational burden.

3   Experimental Results and Analysis

3.1   Experimental Setup

Table 1. Quantitative results of denoising results in simulated experiments.

Noise	Index	Noisy	LRMR	BM4D	LRTV	Proposed
$\sigma = 5$	PSNR	34.585	40.679	41.211	40.543	**41.316**
	SSIM	0.921	0.994	**0.995**	0.993	0.993
	SAM	0.205	0.053	0.059	0.063	**0.049**
$\sigma = 30$	PSNR	17.781	32.074	30.811	29.953	**32.216**
	SSIM	0.362	0.956	0.950	0.941	**0.957**
	SAM	0.779	0.193	0.207	0.181	**0.179**
$\sigma = 50$	PSNR	14.23	28.921	26.785	26.461	**29.074**
	SSIM	0.183	**0.907**	0.898	0.900	0.899
	SAM	0.981	0.192	0.211	0.233	**0.178**
$\sigma = 70$	PSNR	12.133	26.721	25.074	24.839	**26.931**
	SSIM	0.112	**0.892**	0.859	0.881	0.883
	SAM	1.104	0.321	0.368	0.411	**0.318**

To prove the effectiveness of the proposed method, simulated experiments are performed in this paper. Mainstream methods LRMR [6], BM4D [5], LRTV [7] are chosen as baselines to evaluate the proposed model. Each input HSI band is normalized to $[0, 1]$. The adjacent spectral band number K is set to 16. Adam [10] is used to optimize the model with momentum $\beta_1 = 0.9$, $\beta_2 = 0.999$. The learning rate is initialized to 0.0001 and halved every 30 epochs. The Washington DC Mall image obtained by the Hyperspectral Digital Imagery Collection Experiment

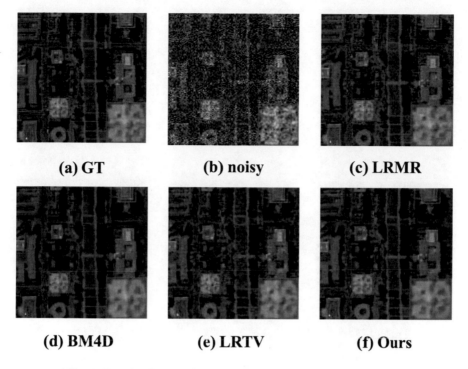

(a) GT   **(b) noisy**   **(c) LRMR**

(d) BM4D   **(e) LRTV**   **(f) Ours**

Fig. 2. Results for the Washington dc Mall image with $\sigma = 50$

Airborne Sensor [11] with the size of $1280 \times 303 \times 191$ is employed to generate the noisy synthetic data. $1080 \times 303 \times 191$ is used for training and the rest part is used for testing. The training dataset is cropped into patches of size 64×64. We use the deep learning framework Pytorch on a machine with NVIDIA GTX 2080Ti GPU, Intel(R) Xeon(R) E5-2678 CPU of 2.50 GHz and 32 GB RAM to train RMRNet.

The simulated noise is generated through imposing additive white Gaussian noise. The noise intensity is multiple and conforms to a fixed distribution or random probability distribution for different experiments. Data augmentation (angles of $0°$, $90°$, $180°$, and $270°$) is employed to increase the number of training samples for better performance.

3.2   Results and Analysis

PSNR, SSIM [12], and SAM [13] are utilized as evaluation metrics in our experiments. Higher PSNR, SSIM, and lower SAM values mean better denoising performance. The averages and standard deviations of the contrasting evaluation indexes of the four cases with different noise levels and are listed in Table 1 and the best performance for each quality index is marked in bold. Under weak noise levels, the BM4D algorithm has a good noise reduction performance, as shown in Table 1 with $\sigma = 5$, but it is not able to deal with strong noise levels such

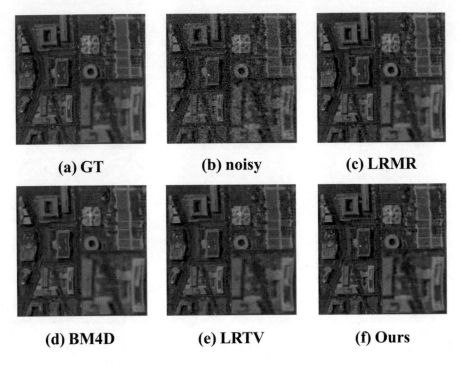

(a) GT   **(b) noisy**   **(c) LRMR**

(d) BM4D   **(e) LRTV**   **(f) Ours**

Fig. 3. Results for the Washington dc Mall image with $\sigma = 30$ in the pseudocolor image with bands (60, 27, 17).

as $\sigma = 100$. LRMR performs well in SSIM under equal noise intensity for different spectra in Table 1. However, it generates some fake artifacts in Fig. 2. The proposed method achieves the highest PSNR and SSIM values and the lowest SAM values in most noise levels. Due to the large number of bands in an HSI, only one band is selected to give the visual results in each case. We show the a single band visual results with $\sigma = 50$ in Fig. 2 and pseudocolor view of bands 17, 27, and 60 in Fig. 2. Our RMRNet is capable of preserving more details and structural information. The noise is eliminated effectively (Fig. 3).

4   Conclusion

In this paper, we proposed a novel RMRNet for HSIs denoising. Both the multiresolution correlation and adjacent spectral information are simultaneously assigned to the proposed network. The experiments indicated that the proposed RMRNet outperforms mainstream models in both evaluation metrics and visual effects.

In our future work, we will investigate more efficient learning framework to remove the mixed noise in HSI data such as stripe noise, impulse noise, deadlines and more complex noisy case. The adaptive methods such as attention mechanism, dynamic convolution and other useful strategies will be taken into consideration in next work.

References

1. Chen, Y., Jiang, H., Li, C., Jia, X., Ghamisi, P.: Deep feature extraction and classification of hyperspectral images based on convolutional neural networks. IEEE Trans. Geosci. Remote Sens. **54**(10), 6232–6251 (2016)
2. Marinelli, D., Bovolo, F., Bruzzone, L.: A novel change detection method for multitemporal hyperspectral images based on binary hyperspectral change vectors. IEEE Trans. Geosci. Remote Sens. **57**(7), 4913–4928 (2019)
3. Buades, A., Coll, B., Morel, J.-M.: Nonlocal image and movie denoising. Int. J. Comput. Vis. **76**(2), 123–139 (2008)
4. Dabov, K., Foi, A., Katkovnik, V., Egiazarian, K.: Image denoising by sparse 3-D transform-domain collaborative filtering. IEEE Trans. Image Process. **16**(8), 2080–2095 (2007)
5. Maggioni, M., Boracchi, G., Foi, A., Egiazarian, K.: Video denoising using separable 4D nonlocal spatiotemporal transforms. In: Proceedings of SPIE, Image Processing: Algorithms and Systems IX, vol. 7870, p. 787003 (2011)
6. Zhang, H., He, W., Zhang, L., Shen, H., Yuan, Q.: Hyperspectral image restoration using low-rank matrix recovery. IEEE Trans. Geosci. Remote Sens. **52**(8), 4729–4743 (2014)
7. He, W., Zhang, H., Zhang, L., Shen, H.: Total-variation-regularized low-rank matrix factorization for hyperspectral image restoration. IEEE Trans. Geosci. Remote Sens. **54**(1), 178–188 (2016)
8. Zhang, Y., Tian, Y., Kong, Y., Zhong, B., Fu, Y.: Residual dense network for image super-resolution. In: Proceedings of the IEEE Conference on Computer Vision and Pattern Recognition (CVPR), June 2018
9. Mao, X., Shen, C., Yang, Y.-B.: Image restoration using very deep convolutional encoder-decoder networks with symmetric skip connections. In: Advances in Neural Information Processing Systems (NeurIPS), pp. 2802–2810 (2016)
10. Kingma, D.P., Ba, J.: Adam: a method for stochastic optimization. In: International Conference on Learning Representations (ICLR) (2015)
11. Wang, Q., Zhang, L., Tong, Q., Zhang, F.: Hyperspectral imagery denoising based on oblique subspace projection. IEEE J. Sel. Top. Appl. Earth Observ. Remote Sens. **7**(6), 2468–2480 (2014)
12. Wang, Z., Bovik, A.C., Sheikh, H.R., Simoncell, E.P.: Image quality assessment: from error visibility to structural similarity. IEEE Trans. Image Process. **13**(4), 600–612 (2004)
13. Yuhas, R.H., Boardman, J.W., Goetz, A.F.: Determination of semiarid landscape endmembers and seasonal trends using convex geometry spectral unmixing techniques. In: Summaries of the 4-th Annual JPL Airborne Geoscience Workshop (1993)

Skin Reflectance Reconstruction Based on the Polynomial Regression Model

Long Ma$^{(\boxtimes)}$ and Yingying Zhu

School of Science, Shenyang Jianzhu University, Shenyang 110168, Liaoning, China

Abstract. Skin spectral reflectance has applications in numerous medical fields including the diagnosis and treatment of cutaneous disorders and the provision of maxillofacial soft tissue prostheses. This paper describes the polynomial model based on the least square (LS) method for skin spectral reflectance from RGB. Furthermore, this paper uses the real human skin data, which makes our results more practical. The performance is evaluated by the mean, maximum and standard deviation of color difference values under other sets of light sources. The values of standard deviation of root mean square (RMS) errors and goodness of fit coefficient (GFC) between the reproduced and the actual spectra were also calculated. Results are compared with the Xiao's method. All metrics show that the proposed method leads to considerable improvements in comparison with the Xiao's method.

Keywords: Skin spectral reflectance reconstruction · Polynomial regression model · Color difference · Root mean square error · Goodness of fit coefficient

1 Introduction

The skin spectral reflectance is gaining importance in many fields, including medical diagnosis [1, 2], computer graphics [3], cosmetic industry [4], and even social sciences [5]. Therefore, it is increasingly important to obtain hyperspectral information of the skin spectral reflectance.

Hyperspectral imaging devices can capture high resolution radiance spectral at every pixel in an image, which is invisible to human eyes and customer RGB cameras [6]. However, the hyperspectral devices are complicated and bulky, which limit their usefulness.

Under any conditions, the spectrum data of a color is unique, which ensures the accuracy of spectral reproduction [7]. Spectral reflectance reconstruction from RGB images is another method to catch hyperspectral information [8–10]. Due to the rapid development of digital cameras in recent years, great progress has been made in image clarity and color reproduction. And the digital camera is easy to carry, flexible to use, and does not need to be contacted for measurement. What's more, digital cameras can measure small areas of skin with high resolution [11]. Therefore, it is easy for us to reconstruct the spectral reflectance and get the RGB images from the digital cameras.

In the past decades, various skin spectral construction algorithms have been proposed. Among them, Imai and his colleagues [12] constructed the skin spectral with principal

© Springer Nature Singapore Pte Ltd. 2021
Y. Wang and W. Song (Eds.): IGTA 2021, CCIS 1480, pp. 10–22, 2021.
https://doi.org/10.1007/978-981-16-7189-0_2

component analysis, and concluded that skin spectral can be reconstructed by 99.5% with the first three basics functions. Chen and Liu [13] proposed a modified Winner estimation for predicting the skin spectral for tissue measurements. Xiao and Zhu [14] proposed a direct method, which transformed camera RGB to skin reflectance directly using a principal component analysis (PCA) approach. The basic function obtained by a real new skin reflectance database. The results show that the new direct method has significantly better performance than the traditional method.

For the employment of the polynomial regression model, several previous literature shows that polynomial regression models can improve results. Connah [15] used the polynomial regression model to reconstruct spectral, which indicated that the polynomial regression model was superior than the standard linear transform. Martinkauppi [16] reconstructed the Arctic Charr's RGB with the polynomial regression, which showed that the result of using the polynomial regression was better. All of these prompt us apply the polynomial regression model to skin spectral reconstruction.

This paper makes two main contributions:

- We use the real human skin data. Because the surface of the skin is uneven, the skin color of each part of the body is not uniform, and the measurement of the skin will be affected by many factors, such as the distance of the measurement, the size of the field of view, the pressure applied to the skin and body area [17] and so on. It is difficult to compare skin measurements obtained with different instrument parameters under different conditions. Therefore, there are still very few real skin data sets at present.
- We propose a new algorithm with the polynomial regression model based on the Xiao's method [14]. We evaluate the performance of the two methods, the proposed method leads to considerable improvements compared with the Xiao's method.

2   Related Work

In this section, the characteristics of the human skin spectral and the polynomial regression model will be introduced.

2.1   Human Skin Spectral

The human skin spectral reflectance has two characteristics [4]. One characteristic is that the spectral reflectance of human skin generally increases gradually. The other characteristics is that skin spectral reflectance shows "W-shaped" or "U-shaped" hollow in the range of 520 nm to 600 nm as shown below which is different from general spectral reflectance (Fig. 1).

2.2   Polynomial Regression Model

Since any function can be approximated by a polynomial, polynomial regression has a wide range of applications [18]. The response vector obtained by the digital camera is appropriately expanded by the polynomial regression model [19]. The polynomial regression expands the channel response to add more channel response information,

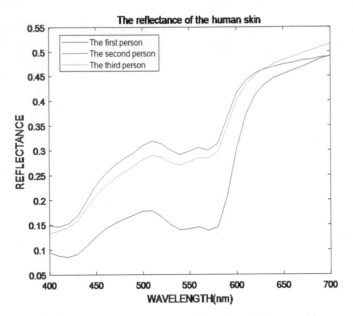

Fig. 1. The spectral reflectance of the three real human skin

thereby improving the accuracy of spectral reconstruction. In this paper, we mainly used the following polynomial regress model, as shown in Table 1. The polynomial can be expanded into second-order (includes 7 items and 10 items), third-order and fourth-order forms. Too many items cause the over-fitting problems [15], we use these usual four expansion forms to expand.

Table 1. Polynomial expansion term

The order of the polynomial (m)	Expansion term
m = 2 (7 items)	1 R G B RG RB GB
m = 2 (10 items)	1 R G B RG RB GB R^2 G^2 B^2
m = 3 (20 items)	1 R G B RG RB GB R^2 G^2 B^2 RGB RG^2 RB^2 GB^2 R^2G R^2B G^2B R^3 G^3 B^3
m = 4 (35 items)	1 R G B RG RB GB R^2 G^2 B^2 RGB RG^2 RB^2 GB^2 R^2G R^2B G^2B R^3 G^3 B^3 R^2GB RG^2B RGB^2 R^2G^2 R^2B^2 G^2B^2 R^3G R^3B G^3B RG^3 RB^3 GB^3 R^4 R^4 R^4

3 Algorithms

In this section, the Xiao's algorithm was reviewed and the proposed algorithm was introduced.

3.1 The Xiao's Algorithm

Xiao et al. [14] proposed an improved method to reconstruct skin reflectance based on principal component analysis (PCA). Skin spectra can be represented by the first k basis functions [20] $B_k = (b_1, b_2, b_3, \ldots, b_k)$, which can be obtained using the new skin database. For each reflectance r, the best coordinates (represented by α) under this set of basic functions are uniquely calculated as the following Eq. (1).

$$\alpha = (B_k)^T r. \tag{1}$$

Where the column vector α represent the best coordinates of the testing sample, B_k represents k basis functions, and T represents the transpose of the matrix. The trained polynomial was used to map the camera RGB to the best coordinate vector α as the Eq. (2).

$$\alpha = \sum_{0 \leq j_1 + j_2 + j_3} a_{j_1, j_2, j_3} R^{j_1} G^{j_2} B^{j_3}. \tag{2}$$

Where $a_{\beta_i, j_1, j_2, j_3}$ is the coefficients of polynomial regression, j_1, j_2 and j_3 are the nonnegative indices, m represents the order of the polynomial regression, and R, G, B are the camera respond signals. Hence, the skin spectral reflectance can be predicted using Eq. (3) as follows.

$$r = B_k \alpha. \tag{3}$$

Where α was derived by the new skin database, the R, G, B were obtained by the polynomial regression model, and the basic functions B_k were derived by the new skin database. The result of reconstruction was improved.

3.2 The Proposed Algorithm

The aim of the skin spectral reconstruction is to recover the spectral reflectance as close to the correct skin spectral reflectance as possible from RGB. We consider that the polynomial model can be trained to map the skin spectral reflectance to the camera RGB, as is shown in Eq. (4).

$$r = \sum_{0 \leq j_1 + j_2 + j_3 \leq m} w_{j_1, j_2, j_3} R^{j_1} G^{j_2} B^{j_3}. \tag{4}$$

Where R, G and B are the camera signals, r is the skin spectra reflectance, m is the order of the polynomial model, and j_1, j_2, j_3 are nonnegative integer indices, and w_{j_1, j_2, j_3} is the model coefficients to be determined. The matrix form of the Eq. (4) is written as:

$$r = M * P. \tag{5}$$

Here, the transformation matrix M can be determined by the training database R_{tr} and P_{tr} is camera response after polynomial regression transformation.

$$R_{tr} = M * P_{tr}. \tag{6}$$

In order to facilitate the solution of M, transpose the two sides of the equation respectively.

$$R_{tr}{}^T = P_{tr}{}^T * M^T. \tag{7}$$

Here, T denotes the transpose of the matrix. $P_{tr}{}^T$ is not a full-rank matrix, and this equation cannot find an accurate solution M. The least-squares regression-based spectral reconstruction seeks the M that minimizes:

$$min\|R_{tr}{}^T - P_{tr}{}^T * M^T\|_F^2. \tag{8}$$

Where $\|R_{tr}{}^T - P_{tr}{}^T * M^T\|_F^2$ denotes the squared F-norm. It follows that we can solve M^T as:

$$M^T = (P_{tr}P_{tr}{}^T)^{-1}P_{tr}R_{tr}{}^T. \tag{9}$$

To transpose r, we get:

$$M = R_{tr}P_{tr}{}^T(P_{tr}{}^TP_{tr})^{-1}. \tag{10}$$

It is solved for, that is. We can use the matrix M, and use the testing sample to reconstruct the skin spectral reflectance as shown in Eq. (11).

$$R_{te} = M * P_{te}. \tag{11}$$

3.3 Regularization

Measurement noise and numerical instability can have very unpleasant effects during the experiment. Due to the noise in the measurement, it may cause the problem of overfitting to the training set. The function of regularization is to use some prior information to reduce the influence of noise in the data, and to provide reasonable estimates where the data is missing or unreliable [21]. Regularization is to add a regular term after the cost function to get the matrix M, as shown in Eq. (12).

Where λ is the regular term coefficient, which is used to weigh the proportion between the regular term and the cost function term.

$$\|R_{tr}{}^T - P_{tr}{}^T * M^T\|_F^2 + \lambda\|M^T\|_F^2. \tag{12}$$

4   Experiment Setup

In this section, the evaluation metrics of the experiment performance and the experiment data will be explained.

4.1   Evaluation Metrics

In this paper, the performance is evaluated by the mean, maximum and standard deviation of color difference values under other sets of light sources. The values of standard deviation of root mean square errors (RMSE) and goodness of fit coefficient (GFC) between the reproduced and the actual spectra were also calculated.

The color differences [22], which can be calculated by the Eq. (13). Where $(\Delta L*)^2$, $(\Delta a*)^2$ and $(\Delta b*)^2$ are the squared differences of the L*a $* b$ transforms of two spectra.

$$\Delta E^*{}_{ab} = \left[\left(\Delta L^*\right)^2 + \left(\Delta a^*\right)^2 + \left(\Delta b^*\right)^2\right]^{\frac{1}{2}} \tag{13}$$

The root-mean-square error shows the two spectral shape differences [23], it is calculated with Eq. (14) as follows. Where $P_m(\lambda_i)$ is the reconstructed spectral reflectance, $P_e(\lambda_i)$ is the true spectral reflectance, and N is the total number of samples.

$$\varepsilon_{RMSE} = \sqrt{\frac{\sum_{i=1}^{N}[P_m(\lambda_i) - P_e(\lambda_i)]^2}{N}} \tag{14}$$

The goodness of fit coefficient can reflect the color difference in perceptual [24], which can be calculated by Eq. (15). Where $P_m(\lambda_i)$ is the reconstructed spectral reflectance, $P_e(\lambda_i)$ is the true spectral reflectance, and N is the total number of samples. hen $\varepsilon_{GFC} \geq 0.99$ is acceptable reconstruction performance, and $\varepsilon_{GFC} \geq 0.999$

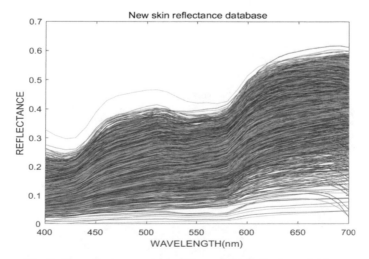

Fig. 2. The reflectance spectra of the new skin reflectance database

is regarded as very good performance and $\varepsilon_{GFC} \geq 0.9999$ means nearly exact reconstruction [6].

$$\varepsilon_{GFC} = \frac{|\sum_{i=1}^{N} P_m(\lambda_i) P_e(\lambda_i)|}{\sqrt{\sum_{i=1}^{N} (P_m(\lambda_i))^2} \sqrt{\sum_{i=1}^{N} (P_e(\lambda_i))^2}}. \tag{15}$$

4.2 Experiment Data

The reflectance spectra of the new skin reflectance database composed of 4392 data [17] shown in Fig. 2, which measured nine body areas of 482 subjects from three ethnic groups (Caucasians, Chinese and Kurdish). The reflectance spectra of 34 test sample [11] data shown in Fig. 3 are measured from real human skin, facial images for 17 subjects were captured and the reflectance of their foreheads and cheeks were measured, the subjects consisted of 8 Caucasians, 8 Chinese and 1 Indian. The reflectance spectra image of 90 silicone skin data [11] shown in Fig. 4 are developed by Spectromatch Ltd. to provide an accurate reference chart for soft tissue prostheses applications, and used silicone to imitate human skin and measured the skin spectral reflectance.

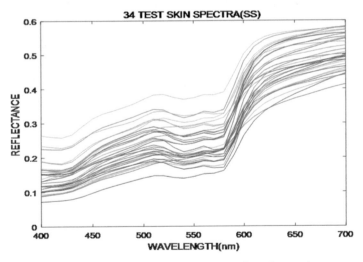

Fig. 3. The reflectance spectra of 34 test Sample samples

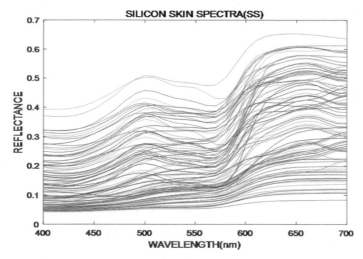

Fig. 4. The reflectance spectra of the 90 silicon skin samples

5 Result and Discussion

The aim of the experiment is to compare the reconstruction accuracy of the Xiao's method and the proposed method. This paper used the Xiao's method and the proposed method to construct the skin spectra reflectance under silicone skin database and the new database (The new database combines test skin spectral and silicone skin database.) respectively. Because the order of the polynomial can influence the accuracy of the reconstruction, we evaluate the reconstruction under the polynomial order m is changed from 2 to 4 and the change of basis vector from 3 to 6 by using leave one out methods (Tables 3 and 4).

Firstly, compared to the reconstruction performance of the Xiao's method in Table 2, the proposed method shows great advantage in spectral reconstruction accuracy.

Next, it can be seen clearly from Table 2 that when the number of basis vectors is 3 and the second order (10 terms) polynomial regression model are used, the Xiao's method can arrive the highest reconstruction accuracy with the silicon skin dataset.

Lastly, for the proposed method, especially when we use the silicon skin database, the reconstruction accuracy increased by 50% at least. Since the two datasets are more similar, the creation accuracy will be higher. Therefore, the silicon skin database is better than the new database.

Table 2. The reconstruction performance of the Xiao's algorithm (the table Only show the best performance of polynomial order from 2–4 when the number of basis functions is constant)

Evaluation procedure	The number of basis function	N = 3	N = 4	N = 5	N = 6
	The polynomial order	2 (10 items)	2 (10 items)	3	3
D65 Color difference	mean	2.0558	2.4475	2.3199	2.3143
	med	1.9144	2.3347	2.217	2.259
	min	0.3601	0.3389	0.1995	0.2832
	max	5.1291	6.2009	6.8947	6.6923
A Color difference	mean	2.179	2.0687	1.8666	1.9136
	med	1.97	1.8484	1.6679	1.7801
	min	0.2411	0.4969	0.3024	0.4636
	max	6.3787	6.8586	8.3483	8.1379
D50 Color difference	mean	2.1305	2.3627	2.2101	2.2052
	med	1.95	2.2005	2.1197	2.1277
	min	0.4838	0.36775	0.2264	0.19359
	max	5.6365	6.5373	7.487	7.2657
F2 Color difference	mean	2.0304	2.273	2.0772	2.1366
	med	1.8474	2.0847	1.9517	2.0244
	min	0.3272	0.46208	0.5013	0.4313
	max	5.4316	5.8918	6.299	6.3538
F11 Color difference	mean	2.6471	2.981	2.7272	2.6297
	med	2.5673	2.8447	2.6017	2.5767
	min	0.4555	0.67105	0.3203	0.6325
	max	7.0536	7.4702	10.053	9.5677
RMSE	mean	0.0158	0.014005	0.0144	0.01428
	med	0.0148	0.013485	0.0132	0.01308
	min	0.0054	0.002158	0.0027	0.00276
	max	0.0571	0.053565	0.0563	0.05607
GFC	mean	0.9983	0.99873	0.9988	0.99887
	med	0.9986	0.99919	0.9993	0.99929
	min	0.9839	0.98616	0.9885	0.98849
	max	0.9994	0.99972	0.99974	0.99973

Table 3. The reconstruction performance of the proposed algorithm under the new database

Evaluation procedure	Sample	The new database			
	The polynomial order	2 (7 items)	2 (10 items)	3	4
D65 Color difference	mean	2.0965	2.084	2.1945	2.3569
	med	1.7017	1.6542	1.7646	1.9386
	min	0.2148	0.2187	0.3146	0.2641
	max	8.1383	8.0423	8.4115	10.408
A Color difference	mean	2.1374	2.1425	2.2624	2.4083
	med	1.6402	1.6227	1.8125	2.0077
	min	0.2663	0.2328	0.2324	0.4387
	max	8.4551	8.262	8.6213	11.407
D50 Color difference	mean	2.1147	2.1077	2.2297	2.3832
	med	1.694	1.6692	1.778	1.9533
	min	0.2381	0.2271	0.3396	0.3087
	max	8.1369	8.0406	8.5615	10.793
F2 Color difference	mean	2.046	2.0313	2.13	2.2866
	med	1.6655	1.5829	1.7251	1.8242
	min	0.3253	0.2235	0.2651	0.0789
	max	8.1532	8.2476	8.4813	10.338
F11 Color difference	mean	2.331	2.349	2.4946	2.6315
	med	1.8672	1.8624	1.9873	2.232
	min	0.5144	0.2963	0.4313	0.3172
	max	8.5804	8.3113	9.3946	12.470
RMSE	mean	0.0178	0.0178	0.0188	0.0203
	med	0.0141	0.0139	0.0153	0.0156
	min	0.0031	0.0028	0.0019	0.0021
	max	0.084	0.083	0.081	0.1127
GFC	mean	0.999	0.999	0.999	0.9991
	med	0.999	0.999	0.999	0.9994
	min	0.990	0.989	0.992	0.9919
	max	0.99996	0.99997	0.99996	0.99998

Table 4. The reconstruction errors of the proposed method under the silicon skin database

Evaluation procedure	Sample	The silicon skin database			
	The polynomial order	2 (7 items)	2 (10 items)	3	4
D65 Color difference	mean	1.4203	1.3146	1.2402	1.4197
	med	1.2861	1.1442	1.0781	1.1897
	min	0.3986	0.2876	0.3341	0.2702
	max	5.9333	5.7321	3.7339	4.5871
A Color difference	mean	1.3946	1.3255	1.2526	1.4371
	med	1.1945	1.091	1.0905	1.2467
	min	0.3534	0.3315	0.2886	0.3162
	max	5.2776	7.2806	4.0333	4.7539
D50 Color difference	mean	1.3974	1.2996	1.2315	1.4153
	med	1.2212	1.1216	1.0632	1.205
	min	0.3760	0.3134	0.3733	0.2928
	max	5.6912	6.3257	3.7914	4.6367
F2 Color difference	mean	1.4075	1.3117	1.2519	1.4739
	med	1.2337	1.0497	1.0612	1.1325
	min	0.3979	0.2622	0.2241	0.2131
	max	6.292	5.3595	3.5224	5.334
F11 Color difference	mean	1.5133	1.444	1.3962	1.6253
	med	1.4001	1.1745	1.2522	1.2587
	min	0.2864	0.1778	0.3165	0.3377
	max	5.5934	7.2449	4.1807	6.1592
RMSE	mean	0.0111	0.0110	0.0116	0.0134
	med	0.0097	0.0096	0.0096	0.0105
	min	0.0027	0.0014	0.0023	0.0022
	max	0.0492	0.0506	0.0545	0.0662
GFC	mean	0.9992	0.9991	0.9993	0.9992
	med	0.9994	0.9995	0.9996	0.9995
	min	0.9927	0.9835	0.9934	0.9943
	max	0.99998	0.99998	0.99998	0.99999

6 Conclusion

Spectral reconstruction is a promising solution to acquiring hyperspectral images, which maps the RGB to hyperspectral images. Skin spectral information is playing an increasingly important role in all walks of life. In order to improve the reconstruction accuracy, this paper uses the polynomial regression model to expand RGB for more channel response information. We evaluation the performance of the proposed method and the Xiao's method by color difference, goodness of fit coefficient, root mean square error. All metrics show that the proposed method leads to considerable improvements in comparison with the Xiao's method. When we use the silicon skin database, the reconstruction accuracy increases by 50% at least.

References

1. Leonardi, A., Buonaccorsi, S., Pellacchia, V., Moricca, L.M., Indrizzi, E., Fini, G.: Maxillo-facial prosthetic rehabilitation using extraoral implants. J. Craniofacial Surg. **19**(2), 398–405 (2008)
2. Nishidate, I., Maeda, T., Niizeki, K., Aizu, Y.: Estimation of melanin and hemoglobin using spectral reflectance images reconstructed from a digital RGB image by the Wiener estimation method. Sensors (Switzerland) **13**(6), 7902–7915 (2013)
3. Zhao, Y., Guo, H., Ma, Z., Cao, X., Yue, T.,Hu, X.: Hyperspectral imaging with random printed mask. In: Proceedings of the IEEE Computer Society Conference on Computer Vision and Pattern Recognition, pp. 10141–10149 (2019)
4. Doi, M., Ohtsuki, R., Tominaga, S.: Spectral estimation of skin color with foundation makeup. In: Kalviainen, H., Parkkinen, J., Kaarna, A. (eds.) Image Analysis. SCIA 2005. Lecture Notes in Computer Science, vol. 3540, pp. 95–104. Springer, Heidelberg (2005). https://doi.org/10.1007/11499145_11
5. Thorstenson, C.: Validation of a method to estimate skin spectral reflectance using a digital camera (2017)
6. Lin, Y.T., Finlayson, G.D.: Physically plausible spectral reconstruction from RGB images. In: IEEE Computer Society Conference on Computer Vision and Pattern Recognition Workshops, pp. 2257–2266 (2020)
7. Li, C., Ronnier Luo, M.: The estimation of spectral reflectances using the smoothness constraint condition. In: Final Program and Proceedings - IS and T/SID Color Imaging Conference, pp. 62–67 (2001)
8. Fu, Y., Zhang, T., Zheng, Y., Zhang, D., Huang, H.: Joint camera spectral sensitivity selection and hyperspectral image recovery. In: Ferrari, V., Hebert, M., Sminchisescu, C., Weiss, Y. (eds.) ECCV 2018. LNCS, vol. 11207, pp. 812–828. Springer, Cham (2018). https://doi.org/10.1007/978-3-030-01219-9_48
9. Shi, Z., Chen, C., Xiong, Z., Liu, D., Wu, F.: HSCNN+: advanced CNN-based hyperspectral recovery from RGB images. In: IEEE Computer Society Conference on Computer Vision and Pattern Recognition Workshops, pp. 1052–1060 (2018)
10. Li, J., Wu, C., Song, R., Li, Y., Liu, F.: Adaptive weighted attention network with camera spectral sensitivity prior for spectral reconstruction from RGB images. In: IEEE Computer Society Conference on Computer Vision and Pattern Recognition Workshops, pp. 1894–1903 (2020)
11. Kamimura, K., Tsumura, N., Nakaguchi, T., Miyake, Y.: Evaluation and analysis for spectral reflectance image of human skin. In: Proceedings of SPIE - The International Society for Optical Engineering, pp. 30–37 (2005)

12. Imai, F.H., Tsumura, N., Haneishi, H., Miyake, Y.: Principal component analysis of skin color and its application to colorimetric color reproduction on CRT display and hardcopy. J. Imaging Sci. Technol. **40**(5), 422–430 (1996)
13. Chen, S., Liu, Q.: Modified Wiener estimation of diffuse reflectance spectra from RGB values by the synthesis of new colors for tissue measurements. J Biomed Opt **17**(3), 030501 (2012)
14. Xiao, K., Zhu, Y., Li, C., Connah, D., Yates, J.M., Wuerger, S.: Improved method for skin reflectance reconstruction from camera images. J. Opt. Express **24**(13), 14934–14950 (2016)
15. Connah, D., Hardeberg, J.Y.: Spectral recovery using polynomial models. In: Proceedings of SPIE - The International Society for Optical Engineering, pp. 65–75 (2005)
16. Martinkauppi, J.B., Shatilova, Y., Kekäläinen, J., Parkkinen, J.: Polynomial regression spectra reconstruction of arctic charr's RGB. In: Trémeau, A., Schettini, R., Tominaga, S. (eds.) Computational Color Imaging CCIW 2009. LNCS (LNAI and LNB), vol. 5646, pp. 198–206. Springer, Heidelberg (2009). https://doi.org/10.1007/978-3-642-03265-3_21
17. Xiao, K., et al.: Characterising the variations in ethnic skin colours: a new calibrated data base for human skin. Skin Res. Technol. **23**(1), 21–29 (2017)
18. Zhu, J., Wen, C., Zhu, J., Zhang, H., Wang, X.: A polynomial algorithm for best-subset selection problem. Proc. Natl. Acad. Sci. U.S.A. **117**(52), 33117–33123 (2021)
19. Hong, G., Luo, M.R., Rhodes, P.A.: A study of digital camera colorimetric characterization based on polynomial modeling. Color Res. Appl. **26**(1), 76–84 (2001)
20. Fubara, B.J., Sedky, M., Dyke, D.: RGB to spectral reconstruction via learned basis functions and weights. In: IEEE Computer Society Conference on Computer Vision and Pattern Recognition Workshops, pp. 1984–1993 (2020)
21. Heikkinen, V., Jetsu, T., Parkkinen, J., Hauta-Kasari, M., Jaaskelainen, T., Lee, S.D.: Regularized learning framework in the estimation of reflectance spectra from camera responses. J. Opt. Soc. Am. A: Opt. Image Sci. Vis. **24**(9), 2673–2683 (2007)
22. Brill, M.H.: Acquisition and reproduction of color images: colorimetric and multispectral approaches. Color Res. Appl. **27**(4), 304 (2002)
23. Heikkinen, V., Jetsu, T., Parkkinen, J., Jääskeläinen, T., Lenz, R.: Estimation of reflectance spectra using multiple illuminations. In: Society for Imaging Science and Technology - 4th European Conference on Colour in Graphics, Imaging, and Vision and 10th International Symposium on Multispectral Colour Science, CGIV 2008/MCS 2008, pp. 272–276 (2008)
24. Liu, Z., Liu, Q., Gao, G.A., Li, C.: Optimized spectral reconstruction based on adaptive training set selection. Opt. Express **25**(11), 12435–12445 (2017)

From Deep Image Decomposition to Single Depth Image Super-Resolution

Lijun Zhao[1]([⊠]), Ke Wang[1], Jinjing Zhang[2], Huihui Bai[3], and Yao Zhao[3]

[1] Taiyuan University of Science and Technology, No. 66 Waliu Road,
Taiyuan 030051, China
leejun@tyust.edu.cn
[2] North University of China, No. 3 Xueyuan Road, Taiyuan 030051, China
[3] Beijing Jiaotong University, Beijing 100044, China
{hhbai,yzhao}@bjtu.edu.cn

Abstract. Although many computer vision tasks such as autonomous driving and robot autonomous navigation as well as object recognition and grasping need to use accurate depth information to improve performance, the captured depth images from practical scene are always troubled by low-resolution and contamination. To resolve this problem, we propose a deep single depth image super-resolution method, which includes three parts: depth dual decomposition block, depth image initialization block and depth image rebuilding block. First, we propose a deep dual decomposition network to separate single low-resolution depth image into two high-resolution parts: fine-detail and coarse-structure images with high quality. Second, weighted fusion mechanism is proposed in depth image rebuilding block for feature integration. Finally, these fused features are fed into residual learning-based reconstruction block in depth image rebuilding block to produce high-quality depth image. Experimental results demonstrate that the proposed method can outperform several state-of-the-art depth map super-resolution methods in term of root mean squared error.

Keywords: Depth image · Image super-resolution · Deep neural network · Image decomposition

1 Introduction

Extremely vital roles are played by depth information in practical applications for a miscellany of computational vision tasks such as the identification and capture of industrial objects or commodity, automatic driving, intelligent robot navigation, etc. Nowadays scene geometric information can be easily captured by depth camera in the consumer class such as Kinect and TOF, and this kind information are always stored as depth images with the limited resolution. The quality of captured depth images is also affected by the complexity of the natural scene and the sensitivity of camera imaging sensors. As a result, only low-resolution

© Springer Nature Singapore Pte Ltd. 2021
Y. Wang and W. Song (Eds.): IGTA 2021, CCIS 1480, pp. 23–34, 2021.
https://doi.org/10.1007/978-981-16-7189-0_3

depth image is available, but its quality cannot meet the practical requirements. Thus, depth image super-resolution (SR) technique should be deeply explored to enhance image quality and increase spatial resolution of depth images [1–5].

The goal of single depth image SR is to cast low-resolution depth image to high-resolution one when only depth image is available. Unlike joint depth super-resolution [1–3], single depth image SR is a more ill-posed yet challenging problem, since only depth information provides geometric structure enhancement without scene's other auxiliary messages. Depth image SR approaches can be coarsely categorized as two classes: classic depth image SR methods and deep learning-based depth image SR methods.

In the early literatures, image filtering is a key way to achieve classic depth image SR methods. For instance, guided image filtering is a kind of fast and non-approximate linear-time filtering, which uses color image as a guidance map to enhance depth image quality [6]. To gradually improve depth accuracy, bilateral filter is iteratively applied into the cost volume for depth image SR [7], since the surfaces of most objects are characterized by piece-wise smooth, and the pixels with similar color values in the same object always have similar depth values. Although image filtering can greatly improve depth quality, these image filtering-based methods always use local information without considering image global relevance. Another way is to construct objective function with certain priors and obtain optimal solution by optimization. In [8], markov random field method is constructed for depth reconstruction by using both depth data potential and depth smoothness prior, whose global solution is achieved by optimization algorithm. To further achieve high quality upsampling, nonlocal means regularization is incorporated into least-square optimization [9] in addition to depth data potential and depth smoothness prior. Meanwhile, a global energy optimization regularized by total generalized variation (TGV) is formulated for depth image SR [10]. Considering that the interdependency exists between color image and depth map, a bimodal co-sparse analysis model is introduced to resolve the inverse problem of depth upsampling [11].

Different from these optimization approaches, sparse depth SR method first segments color image as piece-wise smoothness image to explicitly use color boundary information [12]. After that, corresponding depth regions in each segment are reconstructed respectively and then these depth reconstructions are combined. Unlike all the above methods, patch-based synthesis method searches the best candidate high-resolution patch to match with the given low-resolution depth patch [13]. This method is a new way to achieve depth image SR, but the high-low resolution patch matching procedure is time-consuming. In [5], depth image SR problem is reformulated as the boundary compensation problem, which is solved by multiple residual dictionary learning strategy. Although these classic methods can improve depth accuracy, they cannot be accelerated by the hardwares such as GPU or TPU, which greatly limits its wide usage for computer vision tasks.

Compared with traditional depth image SR methods, a class of deep learning-based image SR approaches have achieved great progresses in the last few years.

These approaches always use deep neural networks to learn nonlinear mapping from low-resolution to high-resolution. For example, super-resolution convolutional neural network (SRCNN) is the earliest representative SR method [14], which solely uses three convolutional layers to play different roles. After that, sparse coding-based network is designed according to sparse coding theory [15], whose topology is interpretable. These two approaches can well resolve natural color image SR problem and they can also be applied for depth image SR. There is a big difference between color image and depth map, that is, color image is used for display, while depth map provides scene geometry information for us. In [16], depth map super-resolution task is cast as a series of novel view synthesis sub-tasks to generate multiple-viewpoint depth maps, which finally form high quality depth image by up-shuffle operation. Without directly learning an end-to-end depth SR mapping, two-stage method [17] firstly uses CNN to predict high-quality edges and then TV synthesis is used to refine low quality depth image with the estimated high-quality edges. In [18], multi-scale network (MS-Net) uses three-step: feature extraction, multi-scale upsampling and reconstruction for depth image SR. The contribution of the above networks lies in the topology. Different from them, a visual appearance-based metric in perceptual deep depth SR method is used as SR loss function to significantly improve 3D object's perceptual quality [19]. Although these methods have improved depth quality, but more accurate depth image is expected to be generated from corresponding low-resolution depth image. Consequently, single depth image super-resolution problem should be further researched.

In this paper, we introduce a deep single depth image super-resolution method, which has three components: depth dual decomposition block, depth image initialization block and depth image rebuilding block. Considering that depth images have salient structures when removing fine detail information, we propose to use a deep dual decomposition network to cast single low-resolution depth image as fine-detail and coarse-structure image with high quality. Meanwhile, depth image SR fusion module is proposed to extract convolutional features from fine-detail image, coarse structure image and initialized image, and then these features are merged together to prepare for the final reconstruction. After feature fusion, these features are fed into residual learning-based reconstruction module of depth image rebuilding block to produce high-quality depth image.

The rest of this paper is organized as follows. First, we will introduce the proposed method in Sect. 2. Secondly, experimental results are given in Sect. 3. Finally, we conclude this paper in Sect. 4.

2   The Proposed Method

Previous depth SR methods always design cascaded convolutional neural networks to learn image nonlinear upsampling mapping from low-resolution to high-resolution. However, this kind of network is always troubled by back-propagation problem, when the total convolutional layer number of deep neural networks is set to be especially large. Although ResNet-like networks can extremely alleviate this problem, existing network topology need to be improved. Meanwhile, most CNN-based methods always only learn direct low-to-high mapping, that is, single-task loss is used to supervise network training. To better train deep neural network, multi-task loss is adopted in the proposed method. An image dual mapping from low-resolution image to high-resolution fine-detail image and coarse-structure image can be learned better than direct low-to-high mapping. Consequently, we design a new CNN structure, which consists of three parts: depth dual decomposition block, depth image initialization block and depth image rebuilding block, as depicted in Fig. 1. Next, these three parts will be detailed.

2.1 Depth Dual Decomposition Block

Low-resolution degraded depth images L can be decomposed as two low-resolution parts: low-resolution coarse-structure image S_{low} and fine-texture image $T_{low} = L - S_{low}$ with traditional image smoothing operators such as L_0 gradient minimization [20–26]. Meanwhile, these operators can be learned by deep CNN, when paired image dataset of $\{L, S_{low}\}$ is available. It is a possible manner to leverage both low-resolution coarse-structure image and fine-texture image for deep depth image SR. However, it is not an elegant manner to improve depth precision in this manner. We propose to learn dual decomposition to predict two high-resolution parts S_{high} and T_{high}, whose input is low-resolution degraded depth images L, because this learning is more useful for depth quality enhancement than simple learning from L to S_{low}. The corresponding labeling images are the coarse-structure S_{GT} and fine-texture image $T_{GT} = I_{GT} - S_{GT}$, which is generated by traditional image smoothing operator of L0 gradient minimization [26]. Here, I_{GT} is the ground-truth image of L. As shown in Fig. 1, depth dual decomposition block is designed to learn a low-to-high mapping from L to a pair of high-resolution images S_{high} and T_{high}. In this block, we use three convolutional layers to extract low-resolution features, after which transposed convolutional layer upsamples these features to obtain F_{2X}^0. Next, F_{2X}^0 and 2X upsampled prior image are concatenated along the channel dimension, whose results F_{2X}^1 are as the input of the next stage. Obviously, these operations from L to F_{2X}^1 can written as: $F_{2X}^1 = \mathcal{F}_{Stage1}(L)$. As done in the first stage, the second stage can be given as follows: $F_{4X}^1 = \mathcal{F}_{Stage2}(F_{2X}^1)$. Here, all the four operations in the second stage consist of three convolutional layers and one transposed convolutional layer. Finally, one convolutional layer is used to cast multi-channel features F_{4X}^1 as a residual map, which are added with 4X-upsampled prior image to get the high-resolution coarse-structure image S_{high}. As done in the high-resolution coarse-structure prediction, the similar network structure is used for high-resolution fine-detail prediction, but the pixel-wise add operation is removed for high-resolution fine-detail prediction.

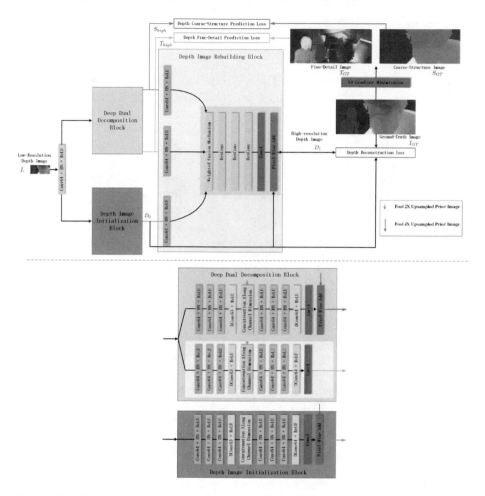

Fig. 1. The diagram of the proposed network. Here, BN denotes the batch normalization, while Conv64 and Conv1 mean that the output channels in these convolutional layers are 64 and 1 respectively.

2.2   Depth Image Initialization Block

Since depth image quality of the final reconstruction in the depth image rebuilding block depends on the quality of triple inputs of this block, e.g., predicted depth coarse-structure image and fine-detail image as well as initialized image, depth image initialization block is introduced to enhance the quality of the low-resolution depth image. As done in the coarse-structure prediction path of depth dual decomposition block, depth image initialization block has the same network topology and they have the same feature maps as their inputs, but they have different outputs. In other words, they have different functions when they learn an end-to-end nonlinear mapping.

2.3   Depth Image Rebuilding Block

To explore potential information of three images: the initialized depth image D_0, predicted depth coarse-structure image S_{high} and fine-detail image T_{high}, three convolutional layers are used to extract shallow features from these three images respectively. Then, we design a weighted fusion mechanism to merge three features from shallow layers together. In this mechanism, the absolute-value is taken for these three features, after which three convolutional layers followed by sigmoid function are utilized to form three weights respectively. Afterwards, these three weights are normalized to get three new weights and then the three shallow features are averaged according to three new weights. To deeply extract the fused feature from the weighted fusion mechanism, we use three ResConvs and one convolutional output layer to extract more informative features and reconstruct the residual image. Finally, the add operation is done to generate the final output D_1 with depth residual image and initialized depth image.

2.4   Loss Function

CNN-based depth image SR methods always only choose single-task loss function to train their network. To better constrain the parameter updating of deep neural network, we optimize the parameter set ζ of depth SR network in the proposed method with the objective function in Eq. (1), in which $|| \cdot ||_1$ denotes L1 norm. Our objective function is composed of two terms: depth dual decomposition loss and depth reconstruction loss. Depth dual decomposition loss includes both depth coarse-structure prediction loss and depth fine-detail prediction loss, which restricts the learning of deep dual decomposition block. Meanwhile, depth reconstruction loss has the initial depth reconstruction loss and the final depth reconstruction loss. Here, the initial depth reconstruction loss supervises the learning of depth image initialization block, while the final depth reconstruction loss affect not only the training of depth image rebuilding block but also the training of depth image initialization block as well as depth dual decomposition block.

$$\arg\min_{\zeta} \underbrace{||T_{high} - T_{GT}||_1 + ||S_{high} - S_{GT}||_1}_{Depth\ Dual\ Decomposition\ Loss} + \underbrace{(||D_0 - I_{GT}||_1 + ||D_1 - I_{GT}||_1)/2}_{Depth\ Reconstruction\ Loss}$$

$$(1)$$

3   Experimental Results

3.1   Training Details

We choose three datasets to form our training dataset. Specifically, the first one is MPI Sintel depth dataset including 58 single depth images. The second one is Middlebury dataset with 34 single depth images, which consists of 6 images from 2001 dataset and 10 images from 2006 dataset, as well as 18 images from 2014 dataset. The third dataset is synthetic training data [13], whose total number

Table 1. Quantitative comparison for depth SR in term of root mean squared error (RMSE) when testing on dataset A.

Image name	Art		Books		Moebius		Ave.
Scalar factor	2X	4X	2X	4X	2X	4X	
Bilinear	2.834	4.147	1.119	1.673	1.016	1.499	2.048
MRFs [8]	3.119	3.794	1.205	1.546	1.187	1.439	2.048
BF [7]	4.066	4.056	1.615	1.701	1.069	1.386	2.316
Park et al. [9]	2.833	3.498	1.088	1.53	1.064	1.349	1.894
GF [6]	2.934	3.788	1.162	1.572	1.095	1.434	1.998
Kiechle et al. [11]	1.246	2.007	0.652	0.918	0.640	0.887	1.058
Ferstl et al. [10]	3.032	3.785	1.290	1.603	1.129	1.458	2.050
SRCNN(RE) [27]	1.614	2.233	1.129	1.106	1.045	1.002	1.355
DnCNN(RE) [28]	0.566	1.619	0.404	0.775	0.354	0.754	0.745
MS-Net*(RE)	1.336	1.941	0.566	0.914	0.588	0.871	1.036
MS-Net(RE)	0.793	1.788	0.439	0.787	0.412	0.766	0.831
MS-Net [18]	0.813	1.627	0.417	0.724	0.413	0.741	0.789
Ours	**0.476**	**1.517**	**0.365**	**0.705**	**0.331**	**0.704**	**0.683**

is 62. Two hole-filled Middlebury RGBD datasets: dataset-A and dataset-C are adopted to evaluate the performance of different depth SR methods. We train the proposed model with Adam optimizer with a learning rate of 2e−4.

3.2 The Objective Quality Comparison of Different Methods

We compare the proposed methods with traditional approaches and CNN-based ones in term of root mean squared error (RMSE). The objective results are provided in Table 1 and 2. Here, SRCNN(RE), MS-Net*(RE), MS-Net(RE) and DnCNN(RE) denote the re-implementation of SRCNN [27], MS-Net [18] and DnCNN [28] according to the network topology described in their papers. MS-Net(RE) uses the residual as input like [18], while MS-Net*(RE) directly feeds low-resolution depth image into corresponding network.

To fairly compare these four methods with the proposed approach, the channel number of convolutional layers in our re-implementation is increased to make the parameter number of these models approximate to that of the proposed method. Note that the result of MS-Net in [18] is also given in Table 1 and 2. From these tables, it can be seen that the proposed method outperforms Bilinear method, MRFs [8], BF [7], Park et al. [9], GF [6], Kiechle et al. [11], Ferstl et al. [10], Lu et al. [12], SRCNN(RE) [27], MS-Net*(RE), MS-Net(RE) and MS-Net [18] for depth 2X and 4X super-resolution. Meanwhile, the RMSE measurement of proposed method is also beyond DnCNN(RE) [28] in average. As described above, our method uses multi-task loss to optimize the proposed network, while depth image initialization block and depth image rebuilding block are designed

Table 2. Quantitative comparison for depth SR in term of root mean squared error (RMSE) when testing on dataset C.

Image name	Tsukuba		Venus		Teddy		Cones		Ave.
Scalar factor	2X	4X	2X	4X	2X	4X	2X	4X	
Aodha et al. [13]	8.993	12.39	2.175	2.597	3.233	4.03	4.262	5.74	5.428
Timofte et al. [27]	9.135	12.09	2.099	2.331	3.253	3.718	4.257	5.49	5.297
Kiechle et al. [11]	3.653	6.212	0.607	0.819	1.198	1.822	1.465	2.974	2.344
Ferstl et al. [10]	5.254	7.352	1.108	1.742	1.694	2.595	2.185	3.498	3.179
Lu et al. [12]	N/A	10.29	N/A	1.734	N/A	2.723	N/A	3.985	N/A
SRCNN(RE) [27]	3.116	7.060	0.772	1.031	1.494	2.013	1.929	3.409	2.603
DnCNN(RE) [28]	1.342	**3.878**	**0.202**	0.436	0.653	1.505	0.646	2.803	1.433
MS-Net*(RE)	3.257	5.954	0.668	0.886	1.220	1.843	1.458	3.063	2.294
MS-Net(RE)	2.186	5.199	0.301	0.762	0.835	1.668	1.005	2.821	1.847
MS-Net [18]	2.472	4.996	0.259	0.422	0.822	1.533	1.100	2.770	1.797
Ours	**1.318**	4.009	0.212	**0.412**	**0.643**	**1.355**	**0.640**	**2.382**	**1.371**

to better reconstruct depth image. As a result, depth accuracy has been greatly improved by the proposed method for depth image SR, when it is compared with the other depth SR approaches.

3.3 The Visual Quality Comparison of Different Methods

In this subsection, we compare the visual quality of different depth map super-resolution methods. In Fig. 2 and Fig. 3, 4X super-resolved images for Moebius depth map from dataset-A and Tsukuba depth map from dataset-C are provided for comparison. From these figures, it can be seen that the SRCNN depth super-resolution method only uses three convolutional layer to extract features, whose receptive field are very limited. In contrast, the edge of depth map MS-Net(RE) and MS-Net*(RE) is more sharper than that of SRCNN, but their quality is less than that of DnCNN (RE), while visual quality of MS-Net(RE) is better than that of MS-Net*(RE). Among all the comparative methods, our method can overcome the above shortcomings, since the proposed method uses multi-task loss functions to supervise the learning of depth map super-resolution network. Additionally, image dual mapping from low-resolution images to high-resolution fine images and coarse structure images is easier than that of direct low-to-high mapping. It can be seen from Fig. 2 and Fig. 3 that the edge of the depth map processed by the proposed method is clearer and the surface is smoother than the other approaches. These visual comparisons further verify the effectiveness of proposed method.

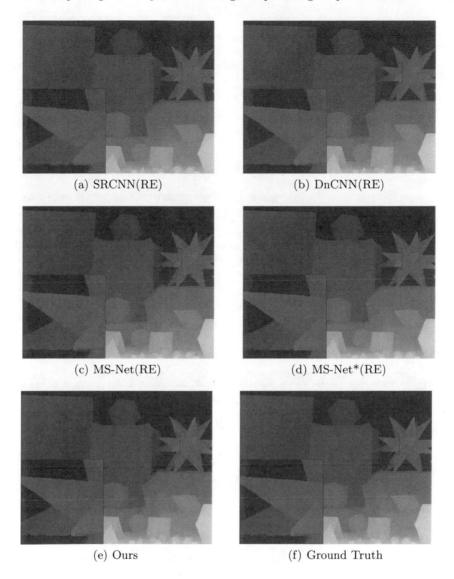

(a) SRCNN(RE)

(b) DnCNN(RE)

(c) MS-Net(RE)

(d) MS-Net*(RE)

(e) Ours

(f) Ground Truth

Fig. 2. The visual comparison of different depth 4X SR approaches for Moebius depth map from dataset-A.

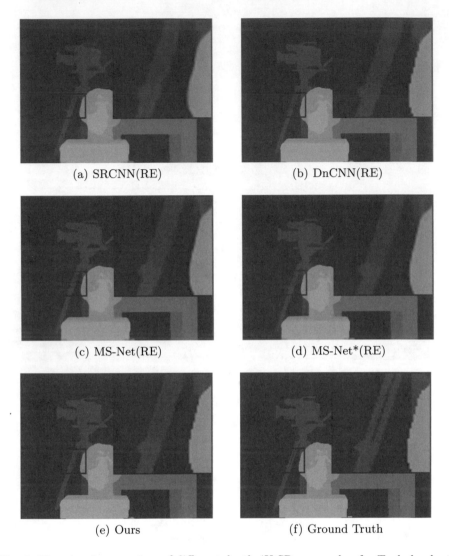

(a) SRCNN(RE)

(b) DnCNN(RE)

(c) MS-Net(RE)

(d) MS-Net*(RE)

(e) Ours

(f) Ground Truth

Fig. 3. The visual comparison of different depth 4X SR approaches for Tsukuba depth map from dataset-C.

4   Conclusion

In this paper, we propose a deep single depth image super-resolution method based on image decomposition, in which deep dual decomposition network is designed to map single low-resolution depth image as high-resolution fine-detail and coarse-structure images with high quality. Meanwhile, both these two images and the initialized image are combined as one group of feature maps according to weighted fusion mechanism. After feature fusion, residual learning-based reconstruction module uses these features in the depth image rebuilding block to produce high-quality depth image. Extensive experimental results show that the proposed method is beyond several traditional and CNN-based depth SR methods. In future work, we will study light-weight deep models for single depth super-resolution based on image decomposition.

Acknowledgments. This work was supported by Doctoral Scientific Research Starting Foundation of Taiyuan University of Science and Technology (No. 20192023), Funding Awards for Outstanding Doctors Volunteering to Work in Shanxi Province (No. 20192055).

References

1. Wang, Z., Ye, X., Sun, B., Yang, J., Xu, R., Li, H.: Depth upsampling based on deep edge-aware learning. Pattern Recogn. **103**(107274), 1–16 (2020)
2. Li, T., Lin, H., Dong, X., Zhang, X.: Depth image super-resolution using correlation-controlled color guidance and multi-scale symmetric network. Pattern Recognit. **107**(107513), 1–13 (2020)
3. Zhao, L., Bai, H., Liang, J., Zeng, B., Wang, A., Zhao, Y.: Simultaneous color-depth super resolution with conditional generative adversarial networks. Pattern Recogn. **88**(1), 356–369 (2019)
4. Zhao, L., Liang, J., Bai, H., Wang, A., Zhao, Y.: Convolutional neural network-based depth image artifact removal. In: IEEE International Conference on Image Processing, Beijing (2017)
5. Zhao, L., Bai, H., Liang, J., Wang, A., Zhao, Y.: Single depth image super-resolution with multiple residual dictionary learning and refinement. In: IEEE International Conference on Multimedia and Expo, Hong Kong (2017)
6. He, K., Sun, J., Tang, X.: Guided image filtering. In: Daniilidis, K., Maragos, P., Paragios, N. (eds.) ECCV 2010. LNCS, vol. 6311, pp. 1–14. Springer, Heidelberg (2010). https://doi.org/10.1007/978-3-642-15549-9_1
7. Yang, Q., Yang, R., Davis, J., Nistr, D.: Spatial-depth super resolution for range images. In: IEEE Conference on Computer Vision and Pattern Recognition, Hawaii (2007)
8. Diebel, J., Thrun, S.: An application of Markov random fields to range sensing. In: Advances in Neural Information Processing Systems, Vancouver (2006)
9. Park, J., Kim, H., Tai, Y.W., Brown, M.S., Kweon, I.: High quality depth map upsampling for 3D-TOF cameras. In: International Conference on Computer Vision, Barcelona (2011)
10. Ferstl, D., Reinbacher, C., Ranftl, R., Rther, M., Bischof, H.: Image guided depth upsampling using anisotropic total generalized variation. In: IEEE International Conference on Computer Vision, Sydney (2013)
11. Kiechle, M., Hawe, S., Kleinsteuber, M.: A joint intensity and depth co-sparse analysis model for depth map super-resolution. In: IEEE International Conference on Computer Vision, Sydney (2013)
12. Lu, J., Forsyth, D.: Sparse depth super resolution. In: IEEE Conference on Computer Vision and Pattern Recognition, Boston (2015)
13. Mac Aodha, O., Campbell, N.D.F., Nair, A., Brostow, G.J.: Patch based synthesis for single depth image super-resolution. In: Fitzgibbon, A., Lazebnik, S., Perona, P., Sato, Y., Schmid, C. (eds.) ECCV 2012. LNCS, vol. 7574, pp. 71–84. Springer, Heidelberg (2012). https://doi.org/10.1007/978-3-642-33712-3_6

14. Dong, C., Loy, C., He, K., Tang, X.: Image super-resolution using deep convolutional networks. IEEE Trans. Pattern Anal. Mach. Intell. **38**(2), 295–307 (2015)
15. Wang, Z., Liu, D., Yang, J., Han, W., Huang, T.: Deep networks for image super-resolution with sparse prior. In: IEEE International Conference on Computer Vision, Santiago (2015)
16. Song, X., Dai, Y., Qin, X.: Deeply supervised depth map super-resolution as novel view synthesis. IEEE Trans. Circuits Syst. Video Technol. **29**(8), 2323–2336 (2018)
17. Chen, B., Jung, C.: Single depth image super-resolution using convolutional neural networks. In: IEEE International Conference on Acoustics, Speech and Signal Processing, Calgary (2018)
18. Hui, T.-W., Loy, C.C., Tang, X.: Depth map super-resolution by deep multi-scale guidance. In: Leibe, B., Matas, J., Sebe, N., Welling, M. (eds.) ECCV 2016. LNCS, vol. 9907, pp. 353–369. Springer, Cham (2016). https://doi.org/10.1007/978-3-319-46487-9_22
19. Voynov, O., et al.: Perceptual deep depth super-resolution. In: IEEE International Conference on Computer Vision, Seoul (2019)
20. Xu, L., Ren, J., Yan, Q., Liao, R., Jia, J.: Deep edge-aware filters. In: International Conference on Machine Learning, Lille (2015)
21. Liu, S., Pan, J., Yang, M.-H.: Learning recursive filters for low-level vision via a hybrid neural network. In: Leibe, B., Matas, J., Sebe, N., Welling, M. (eds.) ECCV 2016. LNCS, vol. 9908, pp. 560–576. Springer, Cham (2016). https://doi.org/10.1007/978-3-319-46493-0_34
22. Li, Y., Huang, J.B., Ahuja, N., Yang, M.H.: Joint image filtering with deep convolutional networks. IEEE Trans. Pattern Anal. Mach. Intell. **41**(8), 1909–1923 (2019)
23. Zhu, F., Liang, Z., Jia, X., Zhang, L., Yu, Y.: A benchmark for edge-preserving image smoothing. IEEE Trans. Image Process. **28**(7), 3556–3570 (2019)
24. Cheng, X., Zeng, M., Liu, X.: Feature-preserving filtering with L0 gradient minimization. Comput. Graph. **38**, 150–157 (2014)
25. Guo, X., Li, S., Li, L., Zhang, J.: Structure-texture decomposition via joint structure discovery and texture smoothing. In: IEEE International Conference on Multimedia and Expo, San Diego (2018)
26. Xu, L., Lu, C., Xu, Y., Jia, J.: Image smoothing via L0 gradient minimization. ACM Trans. Graph. **30**(6), 1–12 (2011)
27. Timofte, R., De Smet, V., Van Gool, L.: Anchored neighborhood regression for fast example-based super-resolution. In: IEEE International Conference on Computer Vision, Sydney (2013)
28. Zhang, K., Zuo, W., Chen, Y., Meng, D., Zhang, L.: Beyond a Gaussian denoiser: residual learning of deep CNN for image denoising. IEEE Trans. Image Process. **26**(7), 3142–3155 (2016)

Classification of Solar Radio Spectrum Based on VGG16 Transfer Learning

Min Chen[1]([✉]), Guowu Yuan[1], Hao Zhou[1], Ruru Cheng[1], Long Xu[2,3], and Chengming Tan[2,3]

[1] School of Information Science and Engineering, Yunnan University, Kunming 650504, China
[2] National Astronomical Observatories, CAS Key Laboratory of Solar Activity, Beijing 100012, China
[3] School of Astronomy and Space Science, University of Chinese Academy of Science, Beijing 100049, China

Abstract. Solar radio bursts are an important part of the study of solar activity, and automatic classification of solar radio spectrum can greatly improve the efficiency of solar activity research. Based on the preprocessing of the original solar radio spectrum images, this paper proposes a solar radio spectrum images classification method based on the VGG16 convolutional neural network and transfer learning. In this method, the pre-trained VGG model is applied to solar radio spectrum recognition. Trained on the generated target data set and adjusted the parameters. The experimental results show that compared with the traditional manual classification method and the existing deep learning classification method, the VGG16 transfer learning classification shows that the TPR of the solar radio burst is better than before. The situation has increased by 12.2%. For the overall classification result analysis, the experimental effect is greatly improved on the basis of the original classification.

Keywords: Classification · Convolutional neural network · Transfer learning · VGG16 model · Denoising · The solar radio spectrum

1 Introduction

The significance of Solar radio spectrum observation is to study solar bursts, and solar radio bursts contained important information about solar activities. There are many types of solar radio bursts that correspond to different physical events. With the development of radio spectrum instruments, the observation data presents a huge trend. It is difficult to detect and classify solar radio bursts manually. Is there a way to efficiently and quickly detect and classify solar radio bursts from this vast amount of information? With deep learning methods have proven to be effective in many complex data classification tasks, deep learning can learn useful features directly from labeled or unlabeled data, which solves many tasks that can only be completed by humans, this provides a way of analyzing and processing solar radio bursts data.

Although some scholars have combined deep learning to classify solar radio spectrum images, they have failed to achieve a good classification effect due to the limited

© Springer Nature Singapore Pte Ltd. 2021
Y. Wang and W. Song (Eds.): IGTA 2021, CCIS 1480, pp. 35–48, 2021.
https://doi.org/10.1007/978-981-16-7189-0_4

amount of data, choice of data form, choice of classification method, the performance of computers, and other reasons. Such as Chen Zhuo, Xu Long, et al. [1, 2] proposed the use of Automatic Encode (AE) [2] and Deep Belief Network (Deep Belief Network). DBN [1] method to learn the characteristics of these massive data and try to use the support vector machine PCA plus SVM [2] method to classify solar radio spectrum data. Chen Sisi et al. [3] proposed the classification algorithm research of solar radio spectrum image based on convolutional neural network, since they are using deep learning methods to automatically learn the representation of solar radio spectrum for the first time, therefore, in the combination classification of solar radio spectrum images and deep learning, there are still some deficiencies in image form selection, image preprocessing and some adjustment of network parameters, resulting in the final classification effect does not reach the expected results.

Because there are too few labeled data of solar radio spectrum and too few training samples for general deep learning algorithms, the classification effect is not good. In this paper, a transfer learning method based on the VGG16 convolutional neural network is proposed to apply the VGG16 network pre-trained in the source domain to the solar radio spectrum images dataset. In this study, we used the data obtained by China's Solar Broadband Radio Spectrometer (SBRS) [4]. After visualizing the original data, image normalization [5] and channel normalization are used to preprocess the image. The processed image samples are divided into a training set and a test set. Fine-tuning in the network to achieve the best parameters of the network, and finally through the test set to verify the model's ability to classify solar radio spectrum images.

2   VGG16 and Transfer Learning

2.1   VGG16 Convolutional Neural Network Model

The VGG16 convolutional neural network model [6] (see Fig. 1) was proposed by the VGG group of the University of Oxford in 2014. It can be known from the development and application of VGG that VGG16 has outstanding achievements in image classification and object detection tasks, which is the reason why this paper chooses the VGG16 convolutional neural network model as the study of transfer learning. VGG16 has a total of 16 layers, 13 convolutional layers, and three fully connected layers. A pooling method will be used after the first two convolutions of 64 convolution kernels; a pooling method will be used after the second two convolutions of 128 convolution kernels; a pooling method will be used after the second two convolutions of 512 convolution kernels; a pooling method will be used after the third fully connections. Multiple convolutional layers and pooling layers are stacked together to make the network have a larger Receptive Field while reducing network parameters. In addition, the original linear function becomes nonlinear through the ReLU activation function, which increases the learning ability of the network. The image is classified by the fully connected layer and the output layer, and the probability distribution of the current sample belonging to different categories can be obtained by the Softmax activation function. The size of the convolution kernel (3 × 3) adopted by this network is to replace the previous larger convolution kernel, and the network can better learn the important features of the image by increasing the depth of the network.

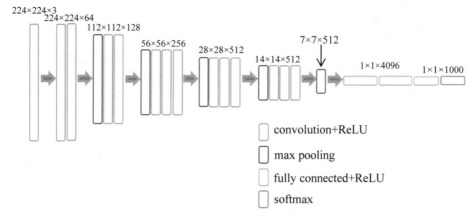

224×224×3
224×224×64
112×112×128
56×56×256
28×28×512
14×14×512
7×7×512
1×1×4096
1×1×1000

☐ convolution+ReLU

☐ max pooling

☐ fully connected+ReLU

☐ softmax

Fig. 1. Structure chart of VGG16 convolutional neural network

2.2 Transfer Learning

With the emergence of more and more machine learning application scenarios, the existing well-performing supervised learning requires a large amount of labeled data. Since labeling data is a boring and costly task, transfer learning has received more and more attention.

Transfer learning [7] is a machine learning method that applies the knowledge learned by convolutional neural networks in other image fields with sufficient labeled data to the research field. This paper introduces the idea of transfer learning because the solar radio spectrum data is limited, and it is not enough to establish a new network structure to train data samples. The overfitting phenomenon is likely to occur in the training process. However, transfer learning can be divided into many categories. According to the transfer scenario, transfer learning can be divided into Inductive Transfer Learning, Transductive Transfer Learning, and Unsupervised Transfer Learning; According to whether the feature space is the same, it can be divided into Homogeneous Transfer Learning and Heterogeneous Transfer Learning; However, the most common transfer method is classified into Instance-based Transfer Learning, Feature-based Transfer Learning, and Parameter-based Transfer Learning. This article mainly uses parameter-based transfer learning, using some network parameters of the pre-trained model VGG16 of Imagenet, which has a huge sample size of natural images and combines solar radio spectrum data with the transfer learning model (see Fig. 2) for training.

3 Spectrum Image Preprocessing and Classification Algorithm

3.1 Preprocessing of Solar Radio Spectrum Data

The solar radio spectrum classification method based on convolutional neural networks is difficult to process the original data directly, so it is particularly important to preprocess the original data. In this paper, the solar radio spectrum data are preprocessed as follows:

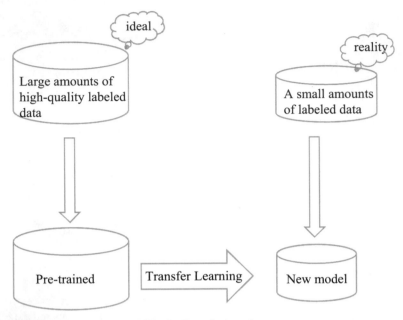

Fig. 2. Transfer learning

(1) Data visualization

The solar radio spectrum data collected by SBRS observation has the characteristics of high time resolution, high-frequency resolution, and high sensitivity, which can produce massive data. The average data collected by observation is about 3-5T per day. The observed and collected solar radio spectrum data are stored in binary form, which can be read in IDL language with HuaiRS software [8–10]. There are many formats of solar radio spectrum data, as shown in Table 1. The data observed by SBRS contains the left and right circular polarization parts of the data (see Fig. 3), because our goal is to accurately classify the three types of images in the solar radio spectrum image of burst, non-burst, and calibration in the data set. To achieve higher accuracy, it is necessary to perform corresponding image processing on the image after data visualization, so that the network can better learn the characteristics of the three types of images.

(2) Channel normalization denoising

In the SBRS data, there are multiple different channels to monitor the solar radio spectrum signals of multiple frequencies at the same time, there is a certain amount of interference between different channels, which causes a large amount of channel noise. However, in the monitoring results, there are little data that originally generated a burst. It is difficult to identify the solar radio spectrum of such a burst. Therefore, the channel normalization method [1] is proposed here to adjust the channel unevenness, reduce the noise in the image, and make the place where the

Table 1. SBRS data format

Frequency	Time	File name/extension	Time resolution
1.0–2.0GHz	1999–2002.5.14	Time information naming	
1.08–2.04GHz	2002.5.14–2002.6.24	.NLP/.NLS/.CLS/.NLM	10 s/0.2 s/0.2 s/5 ms
1.10–2.06GHz	2002.6.25–2004.10.26	.NLP/.NLS/.CLS/.NLM	10 s/0.2 s/0.2 s/5 ms
1.10–1.34GHz	2004.10.24-new	.NLP/.NLS/.CLS/.NLM	10 s/0.2 s/0.2 s/1.25 ms
2.6–3.8GHz	1999-new	.NUP/.NUS/.CUP/.CUS	10 s/0.2 s/0.2 s/8 ms
5.2–7.6GHz	1998.8-new	.NPP/.NUS/.CPS/.NPM	10 s/0.2 s/0.2 s/5 ms

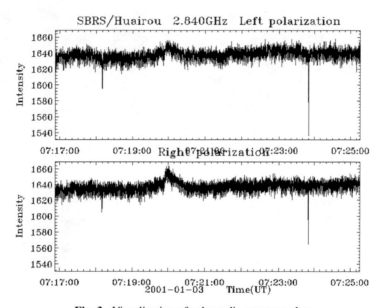

Fig. 3. Visualization of solar radio spectrum data

burst occurs in the image of the explosion more obvious, The formula is as follows:

$$G = f - f_{LM} + f_{GM} \qquad (1)$$

Where f is the visualized image after conversion, G is the image after channel normalization, f_{LM} and f_{GM} represent the local mean and global mean of the image, respectively. f_{LM} is the average value of pixels in a frequency channel, the purpose is to alleviate the influence of horizontal stripe interference caused by channel imbalance. f_{GM} is the average of the pixels of the entire image, the purpose is to add a global to compensate for the background of each pixel. It can be seen from Fig. 4 and Fig. 5 that by using the channel normalization method, the horizontal stripes generated by the channel imbalance can be effectively improved.

Due to the separate visualization of the left circular polarization part (see Fig. 6) and the right circular polarization part (see Fig. 7) of the solar radio spectrum, the radio burst phenomenon is not very obvious. As a result, the introduction of a neural network does not improve classification accuracy. Therefore, the left and right circular polarization (see Fig. 8) are superimposed together in this paper. In this way, when the time is the same and the frequency is different, the phenomenon of solar radio spectrum burst is more obvious, and the classification accuracy of the solar radio spectrum is also improved very well. Visualizations of the solar radio spectrum obtained from the above image processing are as follows: burst (see Fig. 9), non-burst (see Fig. 10), and calibration (see Fig. 11).

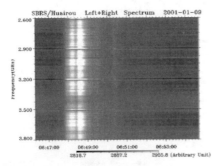

Fig. 4. The solar radio spectrum before channel normalization

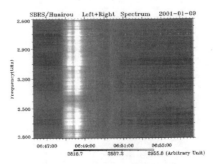

Fig. 5. The solar radio spectrum after channel normalization

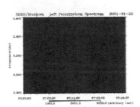

Fig. 6. Left circular polarization spectrum image

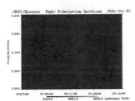

Fig. 7. Right circular polarization spectrum image

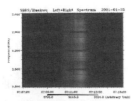

Fig. 8. After left and right circular polarization Superimposed Spectrum

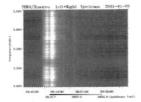

Fig. 9. Burst spectrum image

(3) Image normalization

Since the original visualization graph of data collection is a time-domain signal, there is no obvious difference in the classification of the image. Therefore, the image

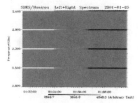
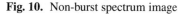

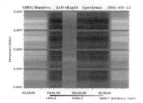

Fig. 10. Non-burst spectrum image   **Fig. 11.** Calibration spectrum image

must be processed to a certain extent to convert the time-domain signal graph into a gray-scale image, under the IDL language to read through the displayed image pattern choice, selected learning network model suitable for depth image mode. The invariant moment of the image is used to find a set of parameters that can eliminate the influence of other transformation functions on image transformation, and the original image to be processed can be converted into the corresponding unique standard form. The image normalization makes the image resistant to the attack of geometric transformation, it can find the invariants in the image, so that these images are originally the same or a series of. The formula for the conversion is as follows:

$$P = \frac{Q - min(Q)}{max(Q) - min(Q)} \tag{2}$$

(4) Multi-scale testing

A visual (798 × 614) grayscale image is obtained after the above data are converted into display mode and the image channel is de-noised. However, the VGG16 network requires the input format of the image to be 224 × 224 × 3 and the pixel to be 224 × 224. Because the new classifier needs to be trained and the weight parameters of the full connected layer are retrained, the size of the source image and the target image must be consistent. The image processing method in this paper is the image preprocessing method in reference [11]. Multi-scale tests (see Fig. 12) were used for image enhancement, re-scale the image with the size of 798 × 614 to 256 × 256, then the shortest side S is 256 at this time. Then random cropping and center cropping are performed from the rescaled training images (each image was cropped once for each SGD iteration). In this way, not only the fixed-size images are sent to the network for training, but also the data enhancement is carried out to increase the stability of the model.

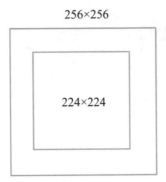

Fig. 12. Multi-scale testing

The total amount of solar radio spectrum data is shown in Table 2:

Table 2. The amount of data marked on the solar radio spectrum

Type	Burst	Non-burst	Calibration	Total
Amount	1846	2060	1614	5520

The total amount of data after data enhancement is shown in Table 3:

Table 3. The amount of data after data enhancement

Type	Burst	Non-burst	Calibration	Total
Amount	9230	10300	8070	27600

3.2 VGG16 Transfer Learning Algorithm

The key to transfer learning is what transfer learning is used, how to perform transfer learning, and when it is suitable for transfer. Since the classification effect of natural images in the classification task of deep learning is the most significant in image classification at present, some network parameters of the pre-trained model VGG16 of the natural image Imagenet with a huge amount of label samples are used for transfer learning. Feature-based Transfer Learning, and Parameter-based Transfer Learning, this article chooses a Parameter-based Transfer Learning, the research of parameter-based migration is to find the common parameters or prior distribution between the spatial model of the source data and the target data, and through further processing, the purpose is to achieve knowledge transfer, the means of transfer learning are:

(1) Transfer Learning: Freeze all the convolution layers of the pre-training model, and only train the self-customized fully connected layer.

(2) Extract Feature Vector: Firstly, the feature vectors of the convolution layer of the pre-training model to all the training and test data are calculated. Then, the pre-training model is abandoned and only the customized fully connected network is trained.

(3) Fine-tuning: Freeze part of the convolutional layer of the pre-training model (usually most of the convolutional layers close to the input, because these layers retain a lot of underlying information) without freezing even any of the network layers, and train the remaining convolutional layers (usually part of the convolutional layers close to the output) and the fully connected layer.

This paper is based on VGG16 for transfer learning, three classifications are achieved through training target data sets. The method is to freeze the feature extraction layer of the original network so that the weights of the convolutional layer and the pooling layer remain unchanged. Since the data to be classified is different from the original VGG16 classification data, the original fully connected layer should be deleted. Two new fully connected layers are added, the classification number of the last fully connected layer matches the number of classes in the data set, and the Softmax function is added to classify the new problems. The parameter information of the last few layers is retrained to achieve the classification goal. Figure 13 shows the process of transfer the model.

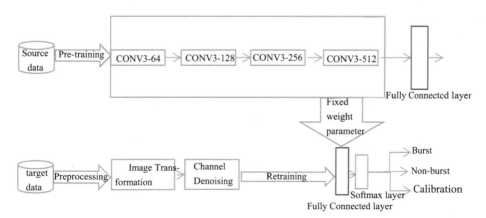

Fig. 13. VGG16 transfer learning process

4 Experimental Results and Analysis

The network is mainly run on the PyTorch framework using Python experiments, using the optimizer is SGD, the learning rate is 0.001, the loss function is cross-entropy, the final classifier is Softmax. The distribution of the training set and test set of solar radio spectrum graph data is shown in Table 4 and Table 5.

Table 4. Solar radio spectrum training set

Type	Burst	Non-burst	Calibration	Total
Amount	7380	8240	6455	22075

Table 5. Solar radio spectrum test set

Type	Burst	Non-burst	Calibration	Total
Amount	1850	2060	1615	5525

In the application process of the VGG transfer model, the parameters are continuously optimized to match the characteristics of the solar radio spectrum data, so that new network layer parameters are trained and then verified with the test set. Due to the lack of solar radio spectrum data, the number of iterations for parameter selection is 10 during the training process, and the number of samples (batch_size) for each gradient descent is 32. Verification is performed after each round and finally calculated by the test set. The accuracy of the classification. From the loss value graph obtained after network training (see Fig. 14), it can be seen that the loss value quickly stabilizes in the iterative process, because the characteristics of the solar radio spectrum image are less complicated compared with the natural image characteristics, its three types (burst, non-burst, calibration) images are grayscale images, and the distribution of features is different so that the network can learn image features better and faster.

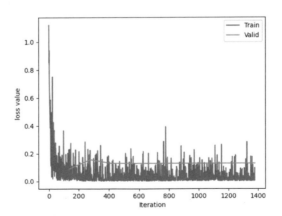

Fig. 14. Visualization result of Loss value

It is important to note that the accuracy here uses TPR (True Positive Rate) and FPR (False Positive Rate) to measure the whole classification. TPR refers to the data that is judged as a positive sample in its environmental test and is still judged as a positive sample after network learning classification. The greater the value of TPR, the better the classification effect of the network model applied. On the contrary, FPR refers to the classification in which the judgment of data in its environmental test is inconsistent with the judgment result after classification in the network model. The greater the value of FPR, the worse the classification effect of the network model applied. The experimental results are shown in Table 6:

Table 6. Statistics of classification results of solar radio spectrum data of VGG16 transfer learning

VGG16 transfer learning		
	TPR (%)	FPR (%)
Burst	96.8	1.4
Non-burst	97.1	1.3
Calibration	99.6	1.8

The experimental results of this paper have achieved a certain improvement in accuracy both compared with the traditional artificial classification method and with the solar radio spectrum classification attempted by many scholars in recent years using deep learning (see Table 7). The burst TPR is 12.2% higher than the previous best case, the non-burst TPR is 7.1% higher than the previous best case, and the calibration TPR is 0.4% lower than the previous best case. The value of FPR is also generally reduced. For example, the value of FPR in the burst is 8% lower than the previous best case, and the value of FPR in the non-burst is 7.4% lower than the previous best case, and the value of FPR in the calibration is 1.1% higher than the previous best case. Although the classification of some images is a little worse than that of previous scholars, in general, the TPR and FPR of the burst are much better than the previous accuracy. However, for the classification of the solar radio spectrum, the improvement of the classification accuracy of the burst data is especially important, because the purpose of this experiment is to classify the burst pictures from a large number of images, and the detected burst parameter values and burst types create indexes together to facilitate query, statistics, and analysis of follow-up research. Forecast the future space weather based on the corresponding analysis results.

Chen Sisi, the author of the literature [3], also used the same VGG16 model to transfer and learn solar radio spectrum data classification. Compared with Chen Sisi's experimental results (see Table 8), the main reasons for the improvement in the classification accuracy of solar radio spectrum data in this paper are as follows:

(1) The form of data visualization is different. The data used by the previous scholars for classification is basically in the form of left circular polarization, and the image phenomenon of the burst after visualization of this data form is not particularly obvious. However, the data sets used in this article are all visualized images after superimposing the left circular polarization and the right circular polarization, so that the explosion phenomenon of the visualized burst image is more obvious, and it also makes the burst image more visually different from the other two images., So that the network can better learn different features to better classify.

Table 7. Comparison of the classification accuracy of the VGG16 transfer learning model and the previous model

	VGG16 transfer learning		CNN		Multimodel		DBN		PCA + SVM	
	TPR (%)	FPR (%)	TPR (%)	FPR (%)	TPR (%)	FPR (%)	TPR (%)	FPR (%)	TPR (%)	FPR (%)
Burst	96.8	1.4	84.6	9.4	70.9	15.6	67.4	3.2	52.7	2.6
Non-burst	97.1	1.3	90	8.7	80.9	13.9	86.4	14.1	0.1	16.6
Calibration	99.6	1.8	100	0.7	96.8	3.2	95.7	0.4	38.3	72.2

Table 8. Chen Si-si used VGG16 transfer learning to classify the accuracy of solar radio spectrum data

Accuracy of solar radio spectrum classification combined with CNN and transfer learning		
	TPR (%)	FPR (%)
Burst	60.1	19.2
Non-burst	64.5	16.6
Calibration	72.8	12.8

(2) Increase in the amount of data. Due to the continuous increase of SBRS statistical data, the burst data obtained also continues to increase, thereby reducing the phenomenon of data imbalance. In Chen Sisi's data set, there are 1158 bursts, 6670 non-bursts, and 988 calibrations. The imbalance of the three types of data can be seen. However, reducing the imbalance of data is also the main reason for the improvement of accuracy.

(3) Different data preprocessing methods. The size of the image processed by Chen Sisi is to process the image into a pixel size of 224 × 224 and send it to the network through continuous dimension upgrade and dimensional reduction. This processing method will cause the image to lose some important feature information before it is sent to the network, leading to network learning If the useful features are reduced, the final classification accuracy will also decrease. The random cropping and center cropping used in this article can not only expand the data, but also establish the weight relationship between each factor feature and the corresponding category, reduce the weight of the background (or noise) factor, and make the model insensitive to missing values. Produce better learning effects, improve model accuracy and increase model stability.

(4) The difference of network parameters and the improvement of operating equipment performance. After the above data preprocessing, combined with the parameters and operating equipment used by the network in the VGG16 transfer learning process,

will have a certain impact on the results. During training, the best results are achieved by continuously updating network parameters suitable for the data set.

5 Conclusion

Aiming at the task of classification of solar radio spectrum data, this paper proposes a classification method based on VGG16 transfer learning. The main tasks are as follows:

(1) Create a new data set. The selected image form of the burst feature is more obvious than other types of images, and the network is easier to recognize the burst feature.
(2) The new method of image preprocessing makes the features of the image clearer and the data-enhanced before the image is sent to the network.
(3) Compared with the traditional manual classification and some existing deep learning model classification, the selected network model for transfer learning achieves a higher classification accuracy.

The experimental results show that by combining the classification of solar radio spectrum data with the deep learning network, the network can also have good feature extraction ability, and the VGG16 transfer learning model designed can also achieve good algorithm performance in the classification task under the condition of limited data.

Acknowledgment. This work is supported by the National Science Foundation of China (Grant No.11663007), the Application and Foundation Project of Yunnan Province (Grant No. 202001BB050032), and the Open Project of CAS Key Laboratory of Solar Activity, National Astronomical Observatories (Grant No. KLSA202115).

References

1. Chen, Z., Ma, L., Xu, L., Tan, C., Yan, Y.: Imaging and representation learning of solar radio spectrums for classification. Multimed. Tools Appl. **75**(5), 2859–2875 (2015). https://doi.org/10.1007/s11042-015-2528-2
2. Zhuo, C., Lin, M., Long, X., et al.: Multimodal Learning for classification of solar radio spectrum. In: 2015 IEEE International Conference on Systems, Man, and Cybernetics. IEEE (2016)
3. Chen, S.-S.: Research on classification algorithm of solar radio spectrum based on convolutional neural network. Shenzhen University (2018)
4. Qijun, F., et al.: A new solar broadband radio spectrometer (SBRS) in China. Solar Phys. **222**(1), 167–173 (2004). https://doi.org/10.1023/B:SOLA.0000036876.14446.dd
5. Zhao, Y.-K., Xu, G.-W., Liu, M.: Bearing fault diagnosis method based on VGG16 transfer learning. Spacecraft Environ. Eng. **37**(05), 446–451 (2020)
6. Zhang, H.-T., Liu, J.-X., Zhao, X.-Q., Hu, X.-H., Li, H.-Y.: Study on VGG16 - based artificial intelligence assisted diagnostic classification of acute lymphoblastic leukemia blood cell microscopic images. Chin. Med. Equip. **34**(07), 1–4+9 (2019)
7. Pan, J., Yang, Q.: A survey on transfer learning. Trans. Knowl. Data Eng. **22**(10), 1345–1359 (2010)

8. Qijun, F., Qin, Z., Ji, H., Pei, L.: A broadband spectrometer for decimeter and microwave radio bursts. Solar Phys. **160**(1), 97–103 (1995). https://doi.org/10.1007/BF00679098
9. Yan, Y., Tan, C., Long, X., Ji, H., Qijun, F., Song, G.: Nonlinear calibration and data processing of the solar radio burst. Sci. China Ser. A: Math. **45**(S1), 89–96 (2002). https://doi.org/10.1007/BF02889689
10. Tan, C., Yan, Y., Tan, B., Qijun, F., Liu, Y., Guirong, X.: Study of calibration of solar radio spectrometers and the quiet-sun radio emission. Astrophys. J. **808**(1), 61 (2015)
11. Simonyan, K., Zisserman, A.: Very deep convolutional networks for large-scale image recognition. Computer Science (2014)

A Channel Attention-Based Convolutional Neural Network for Intra Chroma Prediction of H.266

Yao Liu[1], Xin Ma[1(✉)], Hui Yuan[2], Ye Yang[2], and Qi Liu[1]

[1] School of Information Science and Engineering, Shandong University, Qingdao, Shandong, China
{liuyao903,sdql}@mail.sdu.edu.cn, max@sdu.edu.cn
[2] School of Control Science and Engineering, Shandong University, Jinan, Shandong, China
huiyuan@sdu.edu.cn, yangye@mail.sdu.edu.cn

Abstract. Chroma intra prediction is an important module in Versatile Video Coding (VVC) and Cross-Component Linear Model (CCLM) is an effective coding tool for it, which establish a linear model between the predicted chroma component and the reconstructed luma component. When the video content has complex textures, the chroma prediction performance of CCLM will be suppressed. To further improve chroma prediction ability, we present a simple yet efficient channel attention-based network to predict chroma component, namely, CACNN. The proposed channel attention module is significant, which can control the contribution of each neighboring reference sample when predict chroma component in the current block. We also use multi-line reference samples to further improve the chroma prediction performance. The proposed CACNN is incorporated into the VVC test model version 8.2 (VTM 8.2). Experimental results demonstrate that comparing with VTM 8.2 anchor, the proposed method can achieve 2.89%, 2.36% chroma components bit rate savings in high QPs.

Keywords: Versatile video coding · Chroma intra prediction · Channel attention

1 Introduction

Video is an important information carrier. In recent years, with the gradual increase in the application fields of video, such as entertainment video, telemedicine, surveillance video, teleconference, etc., the number of videos has also increased sharply. At the same time, people have higher requirements for video quality, such as High Definition (HD), High Dynamic Range (HDR) [1] and Wide Color Gamut (WCG) [1] that have emerged over the past few years. Due to the increase in the number of videos and the improvement in video quality, the amount of video data has greatly increased, which has brought great challenges to the storage and transmission of videos. Therefore, improving video compression performance is an urgent problem. At present, the latest International Video Coding Standard is Versatile Video Coding (VVC) [2].

© Springer Nature Singapore Pte Ltd. 2021
Y. Wang and W. Song (Eds.): IGTA 2021, CCIS 1480, pp. 49–59, 2021.
https://doi.org/10.1007/978-981-16-7189-0_5

Compared with HEVC [3], the previous generation of International Video Coding Standard, the coding framework of VVC does not improved. It still adopts a block-based hybrid coding framework, which mainly includes intra prediction coding module, inter prediction coding module, transform coding module, quantization coding module and entropy coding module. But almost all the modules in VVC have been improved with new coding tools. In particular, the intra prediction module includes luma intra prediction and chroma intra prediction. The luma intra prediction mode of VVC has been increased from 35 to 67 compared with HEVC, where 32 more angular modes are included to adapt to the diverse contents. Different from luma intra prediction, there are 8 chroma intra prediction modes in VVC, which are Planar, DC, Vertical, Horizontal, Derived Mode (DM), Cross Component Linear Model (CCLM) and Multi-Directional Linear Model (MDLM), where MDLM consists of left (MDLM_L) and top (MDLM_T) versions. Among the 8 chroma intra prediction modes, CCLM, MDLM_L, and MDLM_T are unique to VVC. When using CCLM, MDLM_L, and MDLM_T, the chroma component is predicted from the already-reconstructed luma samples using a linear model, the two parameters of the linear model are derived from the neighboring reconstructed luma and chroma component. Although the linear models have achieved coding gain to a certain degree, there are still several shortcomings. Firstly, when the content of the prediction block is complex, a single linear model cannot accurately describe the relationship between the reconstructed luma component and the chroma component to be predicted. Secondly, the linear models are obtained by manual design, which may limit chroma prediction performance. To alleviate the above issues, the deep learning-based methods have appeared and shown impressive chroma prediction performance.

In [4], a hybrid network was proposed to predict chroma components, which first extracts the features from the reconstructed luma samples of the current block. Then, using a fully connected network to extract features from the neighboring reconstructed luma and chroma samples. Finally, the extracted twofold features are fused to predict the Cb and Cr component. It is worth mentioning that they use the same network but with different hyperparameters for 4×4, 8×8 and 16×16 chroma blocks. [5] proposed a novel network for chroma intra prediction, which proposed an efficient attention mechanism to modulate spatial relations between reference and predicted samples. Similar to [4], [5] also used the same network structure but with different hyperparameters for 4×4, 8×8 and 16×16 chroma blocks. Zhu et al. [6] presented a deep learning-based intra chroma prediction method (CNNCP), which included two sub-networks for luma down-sampling and chroma prediction. Different from [4, 5], one CNNCP model is only applied for 64×64 chroma block, such that the chroma blocks smaller than 64×64 will copy the chroma predictions of the corresponding position. Compared with [4, 5], [6] is not only more efficient, but also can further improve the coding performance. To further reduce the redundancy between luma component and chroma component, we propose a novel network for intra chroma prediction. Our main contributions include:

(1) We first use the channel attention mechanism to better explore the relationship between the neighboring reference sample and the predicted chroma component.
(2) We use multi-line reference samples to improve the accuracy of chroma prediction.

The remainder of this paper is structured as follows. Section 2 introduces the proposed network CACNN. The experimental results and analyses are provided in Sect. 3. Finally, we conclude in Sect. 4.

2 Proposed Method

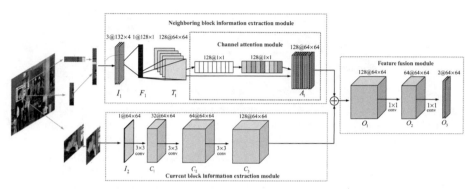

Fig. 1. Framework of the proposed channel attention-based CNN.

The proposed channel attention-based CNN (CACNN) aims to predict the chroma component of the current block by using the neighboring luma and chroma components and the current already reconstructed luma component. As show in Fig. 1, the proposed CACNN includes three modules: neighboring block information extraction module, current block information extraction module and feature fusion module. The neighboring block information extraction module aims to explore the relationship between the neighboring reference sample and the predicted chroma components. The current block information extraction module extracts features from the current reconstructed luma block. Then, the features from the neighboring block information extraction module and current block information extraction module are added at corresponding positions to achieve feature fusion. Finally, the fused feature is fed into feature fusion module to predict the final chroma components (i.e., Cb and Cr).

It is worth noting that our network is used to predict chroma component of 64 × 64 chroma block and is suitable for YUV4:2:0 format. In the next, we will discuss the detailed structure of neighbor information extraction module, current block information extraction module and feature fusion module.

2.1 Neighboring Block Information Extraction Module

Neighboring block information extraction module extracts weighted features from neighboring reference Y, Cb and Cr components. As shown in Fig. 2, The neighboring reference Y is composed of 8 rows of neighboring reconstructed Y component adjacent to the top of the current reconstructed Y block and 8 columns of neighboring reconstructed Y

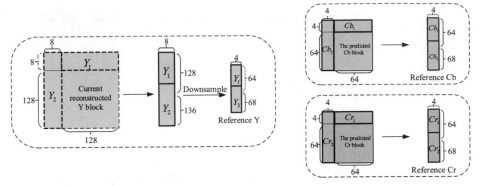

Fig. 2. Schematic diagram of obtaining neighboring reference samples

component adjacent to the left of the current reconstructed Y block and then it is down-sampled by a factor of 2. The neighboring chroma reference component is composed of 4 rows of neighboring reconstructed chroma component adjacent to the top of the current predicted chroma block and 4 columns of neighboring reconstructed chroma component adjacent to the left of the current predicted chroma block. By concatenating the neighboring reference Y, Cb and Cr components, we obtain the input of neighboring block information extraction module, denoted as $\mathbf{I}_1 \in \mathbb{R}^{3 \times 132 \times 4}$. $\mathbf{I}_1$ is fed into a fully connected layer and we can obtain features $\mathbf{F}_1(\mathbf{I}_1) = [f^1, f^2, \cdots, f^{128}]$, $f^i \in \mathbb{R}$. Then, We tile $\mathbf{F}_1(\mathbf{I}_1) = [f^1, f^2, \cdots, f^{128}]$ into the matrix $\mathbf{T}_1(\mathbf{I}_1) = [\mathbf{t}^1, \mathbf{t}^2, \cdots, \mathbf{t}^{128}]$, $\mathbf{t}^i \in \mathbb{R}^{1 \times 64 \times 64}$.

$$t^i_{x,y} = f^i, i \in [1, 128], x, y \in [1, 64] \tag{1}$$

Finally, the weighted features $\mathbf{A}_1(\mathbf{I}_1) \in \mathbb{R}^{128 \times 64 \times 64}$ are produced using the channel attention module [7]. The details of the channel attention module [7] is shown in Fig. 3, $\mathbf{T}_1(\mathbf{I}_1) = [\mathbf{t}^1, \mathbf{t}^2, \cdots, \mathbf{t}^{128}]$ is first fed into the global average pooling layer and go through two fully-connected (FC) layers, then the weighted score $\mathbf{V}_1(\mathbf{I}_1) = [v^1, v^2, \cdots, v^{128}]$ can be obtained by a sigmoid activation.

We denoted the final output of the channel attention module as $\mathbf{A}_1(\mathbf{I}_1) = [\mathbf{a}_1, \mathbf{a}_2, \cdots, \mathbf{a}_{128}]$, $\mathbf{A}_1(\mathbf{I}_1)$ is obtained by rescaling $\mathbf{T}_1(\mathbf{I}_1)$ with the weighted score $\mathbf{V}_1(\mathbf{I}_1)$:

$$\mathbf{a}_i = \mathrm{F}_{scalar}(\mathbf{t}^i, v^i) = \mathbf{t}_i v^i, i \in [1, 128] \tag{2}$$

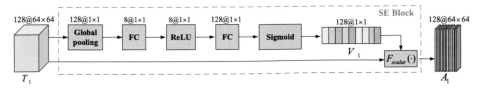

Fig. 3. Channel attention module

2.2   Current Block Information Extraction Module

The current block information extraction module aims to extracts features from the current reconstructed luma block. The input of the current block information extraction module is the reconstructed luma samples $\mathbf{I}_2 \in \mathbb{R}^{1 \times 64 \times 64}$. Then consecutive 3×3 convolutional layers are applied to extracts current reconstructed luma block features. In this process, we get features $\mathbf{C}_1(\mathbf{I}_2) \in \mathbb{R}^{32 \times 64 \times 64}$, $\mathbf{C}_2(\mathbf{I}_2) \in \mathbb{R}^{64 \times 64 \times 64}$ and $\mathbf{C}_3(\mathbf{I}_2) \in \mathbb{R}^{128 \times 64 \times 64}$. $\mathbf{C}_1(\mathbf{I}_2), \mathbf{C}_2(\mathbf{I}_2), \mathbf{C}_3(\mathbf{I}_2)$ are represented as

$$
\begin{cases}
\mathbf{C}_1(\mathbf{I}_2) = \sigma(\mathbf{W}_1 * \mathbf{I}_2) \\
\mathbf{C}_2(\mathbf{I}_2) = \sigma(\mathbf{W}_2 * \mathbf{C}_1(\mathbf{I}_2)) \\
\mathbf{C}_3(\mathbf{I}_2) = \sigma(\mathbf{W}_3 * \mathbf{C}_2(\mathbf{I}_2))
\end{cases}
\tag{3}
$$

where $\mathbf{C}_1(\mathbf{I}_2), \mathbf{C}_2(\mathbf{I}_2), \mathbf{C}_3(\mathbf{I}_2)$ are the extracted features, σ refers to the ReLU function [8], $\mathbf{W}_1, \mathbf{W}_2, \mathbf{W}_3$ are 3×3 convolutional filters.

2.3   Feature Fusion Module

The feature fusion branch is used to map the fused features from the neighboring block information extraction module and the current block information extraction module into the final output Cb and Cr predictions. The fused features $\mathbf{O}_1(\mathbf{I}_1, \mathbf{I}_2)$ is produced by adding the output of the neighboring block information extraction module $\mathbf{A}_1(\mathbf{I}_1)$ and the output of the current block information extraction module $\mathbf{C}_3(\mathbf{I}_2)$ at the corresponding positions.

$$
\mathbf{O}_1(\mathbf{I}_1, \mathbf{I}_2) = \mathbf{A}_1(\mathbf{I}_1) + \mathbf{C}_3(\mathbf{I}_2)
\tag{4}
$$

Finally, the Cb and Cr predictions, denoted as $\mathbf{O}_3(\mathbf{I}_1, \mathbf{I}_2)$, are produced using the consecutive 1×1 convolutional operations:

$$
\begin{cases}
\mathbf{O}_2(\mathbf{I}_1, \mathbf{I}_2) = \sigma(\mathbf{W}_4 * \mathbf{O}_1(\mathbf{I}_1, \mathbf{I}_2)) \\
\mathbf{O}_3(\mathbf{I}_1, \mathbf{I}_2) = \sigma(\mathbf{W}_5 * \mathbf{O}_2(\mathbf{I}_1, \mathbf{I}_2))
\end{cases}
\tag{5}
$$

Where $\mathbf{O}_2$ are the extracted features, σ refers to the ReLU function, $\mathbf{W}_4, \mathbf{W}_5$ are 1×1 convolutional filters.

3   Experiment Results

Our training dataset consisted of two parts. The first part includes 800 images from DIV2K database which are used for training. The second part includes 1124 images from Youku. In detail, we sought 1124 videos from Youku and extracted the first frame of the video to get 1124 images. Finally, the 1924 images are encoded by the VVC Test Model (VTM) version 8.2 [9] with QP = 22 under all intra (AI) configuration.

For training phase, to enhance the generalization ability of the network, we adopted a simple data augmentation strategy, i.e., randomly select the patch with a size of 128×128 as the input of the CACNN. The initial learning rate is 1×10^{-4}. We implemented our method on the Pytorch platform with a single NVIDIA RTX 2080 Ti GPU.

Next, we will briefly introduce how to integrate the CACNN into VVC test model VTM 8.2. Similarly to [6], the CACNN was integrated in both the video encoder and decoder and we took the CACNN as the 9th chroma intra prediction mode, namely, the CACNN mode. At the encoder side, optimal chroma intra prediction mode selection is conducted based on rate distortion optimization (RDO), such that the mode with the minimum RD cost will be selected. An additional binary flag is used to indicate that the current block choose the traditional chroma intra prediction mode or the CACNN mode. At the decoder side, the binary flag will be decoded firstly. If the binary flag is 1, the optimal chroma intra prediction mode of the current block is CACNN mode, else is the traditional chroma intra prediction mode.

3.1   The Robustness Analysis of Chroma Prediction Performance

In order to verify the accuracy of the proposed network for predicting the chroma component of patches encoded by different QPs, we selected VVC test sequences to evaluate the performance of the proposed network. We selected the first frame of each VVC test sequence, and use QPs {22, 42, 52} to encode these images. For each encoded image, we intercepted some 256×256 patches from it with a step size of 128. Finally, for each image in class A, B, C and E, we take out 420, 91, 32, 10 patches, respectively. All patches are fed into CACNN for chroma prediction. The average of the PSNR of all patches in an image is used as the PSNR of the image. The quantitative results were shown in Table 1 (PSNR was used to assess the prediction performance). The CACNN trained by QP22 is applied to predict the chroma component of the patch encoded with QPs {22, 42, 52}, the average Cb PSNR values are 34.68dB, 34.43 dB and 33.76 dB; the average Cr PSNR values are 34.38 dB, 34.13 dB and 33.24 dB, respectively. It can be seen that the coding performance of CACNN did not does not fluctuate significantly under other unseen QPs (i.e., 42 and 52). In other words, CACNN showed strong generalization ability.

Table 1. Prediction performance evaluation with different QPs (PSNR).

Class	Sequence	QP22		QP42		QP52	
		Cb	Cr	Cb	Cr	Cb	Cr
A	Tango2	43.33	38.92	42.64	38.42	39.97	36.49
	FoodMarket4	39.71	39.82	39.40	39.42	38.11	37.21
	Campfire	38.88	35.51	37.18	34.95	36.34	33.50
	CatRobot1	34.11	32.89	34.00	32.78	33.32	31.87
	DaylightRoad2	40.43	37.99	40.14	37.71	39.19	36.10
	ParkRunning3	30.30	33.01	30.24	32.94	29.64	32.48
B	MarketPlace	37.41	37.81	37.39	37.46	37.02	36.60
	RitualDance	40.97	39.84	40.96	39.34	40.58	36.97

(*continued*)

Table 1. (*continued*)

Class	Sequence	QP22		QP42		QP52	
		Cb	Cr	Cb	Cr	Cb	Cr
	Cactus	32.14	30.44	32.14	30.44	31.71	30.07
	BasketballDrive	36.09	34.74	35.90	34.68	34.65	33.71
	BQTerrace	35.81	37.82	35.69	37.40	35.27	37.34
C	BasketballDrill	29.41	31.87	29.41	31.96	29.29	31.57
	BQMall	30.84	30.34	30.76	30.31	30.81	30.46
	PartyScene	27.95	27.28	27.93	27.35	27.79	27.29
	RaceHorses	25.31	26.30	25.24	26.20	24.72	25.48
E	FourPeople	34.83	34.85	34.65	34.65	34.73	34.09
	Johnny	34.31	35.22	33.98	34.61	33.22	34.39
	KristenAndSara	32.51	34.29	32.17	33.84	31.45	32.77
Average		34.68	34.38	34.43	34.13	33.76	33.24

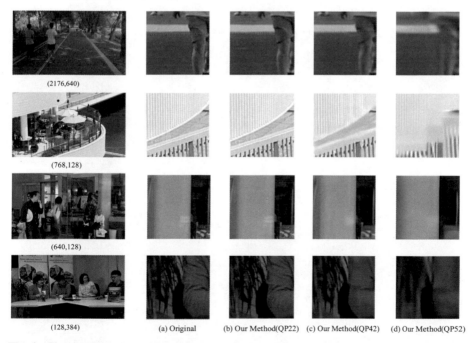

(2176,640)

(768,128)

(640,128)

(128,384)   (a) Original   (b) Our Method(QP22)   (c) Our Method(QP42)   (d) Our Method(QP52)

Fig. 4. Chroma prediction results of our method under different QP, (x, y) represents the position coordinates of the upper left corner of the patch.

To visualize the prediction results of our network under different QPs, we selected the first frame of the four sequences of ParkRunning3, BQTerrace, BQMall and FourPeople from Class A, B, C, E, and encoded them with QPs {22,42,52}. For each image encoded by different QPs, selecting a patch with a size of 128×128 to visualize the prediction results. Figure 4 shown the qualitative results (from top to bottom, the sequence indicates ParkRunning3, BQTerrace, BQMall, and FourPeople, respectively).

3.2   Coding Performance

The proposed method was tested under the common test conditions (CTC) [10], test sequences included 18 video sequences known as Classes A, B, C and E. Since our training set was all natural sequences and the proposed method was suitable for large-resolution videos, the screen content sequences (Class F) and small resolution video (Class D) were removed from the test sequences. We use the all-intra configuration and QPs were set to 42, 47, 52 and 57, respectively. The results are summarized in Table 2. It can be observed that compared with VTM 8.2, an average 2.89% and 2.36% BD-rate

Table 2. BD-rate results anchoring to VTM version 8.2 (Our method).

Class	Sequence	Cb	Cr
A	Tango2	−14.08%	−4.12%
	FoodMarket4	4.27%	−1.90%
	Campfire	3.07%	−0.47%
	CatRobot1	−3.92%	−3.15%
	DaylightRoad2	−2.93%	−14.18%
	ParkRunning3	−3.02%	−3.34%
B	MarketPlace	−2.27%	−1.07%
	RitualDance	−9.57%	1.54%
	Cactus	−2.20%	−4.42%
	BasketballDrive	−4.41%	−0.50%
	BQTerrace	−7.65%	3.79%
C	BasketballDrill	−2.68%	−1.87%
	BQMall	−2.82%	−1.59%
	PartyScene	−3.56%	−3.29%
	RaceHorses	7.10%	5.81%
E	FourPeople	−6.67%	−4.83%
	Johnny	0.43%	−8.51%
	KristenAndSara	−1.10%	−0.42%
Average		−2.89%	−2.36%

Table 3. BD-rate results anchoring to VTM version 8.2 [6].

Class	Cb	Cr
A	3.82%	−2.58%
B	0.06%	2.40%
C	0.21%	1.93%
E	−0.09%	−2.12%
Overall	0.68%	0.31%

[11] reductions can be achieved by the proposed method for Cb and Cr components, respectively.

Additionally, in order to compare with the method proposed in [6], under the premise of ensuring that all configurations (dataset, loss functions, etc.) remain unchanged, we retrain the network model proposed in [6] and use the retrained network model to replace our proposed network model in VTM 8.2. As shown in Table 3, the final BD-rate increase by 0.68% and 0.31% for Cb and Cr components, respectively, which indicates that the proposed method achieves better performance.

To further illustrate the effectiveness of the proposed method, we have counted the proportion of blocks for which the proposed method is selected for chroma intra prediction under different QPs for each test sequence. The results were shown in Table 4. It can be seen that the percentage that selects CACNN can reach 13.4%, 21.7%, 31.7%, and 42.7% on average for four different QP settings. Meanwhile, we observed that with the increase of QP, more and more blocks were chose to use the proposed method, which meant that our method may be more effective for low quality video.

Finally, as shown in Fig. 5, we provide the visualization results of the chroma block division on the sequence ParkRunning3 (3840 × 2160), BQTerrace (1920 × 1080), BQMall (832 × 480), and FourPeople (1280 × 720), where the red block represents that the block selects the CACNN mode for chroma intra prediction. They are encoded with all-intra configuration and QP was set to 42. It also can be seen from Table 4 that in this configuration, 5.2%, 18.5%, 11.4%, and 15.6% of the chroma blocks for the ParkRunning3, BQTerrace, BQMall, and FourPeople sequences choose to use the CACNN mode.

Through the visual block results, we can also see that it was easier to use CACNN for chroma intra prediction in regions with complex texture, it further proved that CACNN can fully explore the relationship between the neighboring reference sample and the current predicted chroma components.

(a) ParkRunning3   (b) BQTerrace

(c) FourPeople   (d) BQMall

Fig. 5. CACNN selected in chroma intra prediction.

Table 4. Percentage of the proposed method selected.

Class	Sequence	QP			
		42	47	52	57
A	Tango2	19.20%	29.50%	35.70%	44.50%
	FoodMarket4	11.40%	20.70%	38.10%	46.20%
	Campfire	15.60%	25.50%	39.10%	49.10%
	CatRobot1	12.20%	21.30%	33.90%	46.80%
	DaylightRoad2	24.70%	36.50%	54.70%	65.00%
	ParkRunning3	25.20%	28.40%	37.40%	37.50%
B	MarketPlace	10.80%	16.00%	28.10%	54.20%
	RitualDance	10.10%	23.90%	37.40%	47.00%
	Cactus	13.80%	17.30%	30.70%	48.10%
	BasketballDrive	13.90%	19.30%	32.00%	48.00%
	BQTerrace	18.50%	29.50%	47.40%	59.80%
C	BasketballDrill	9.00%	13.20%	13.60%	31.70%
	BQMall	11.40%	15.20%	16.80%	28.80%
	PartyScene	8.00%	16.30%	22.50%	14.20%
	RaceHorses	7.00%	13.60%	13.10%	22.80%
E	FourPeople	15.60%	29.00%	38.30%	57.90%
	Johnny	10.40%	21.60%	27.80%	37.90%
	KristenAndSara	5.00%	14.40%	24.10%	29.10%
Average		13.40%	21.70%	31.70%	42.70%

4   Conclusion

In this paper, we proposed a channel attention-based CNN (i.e., CACNN) for intra chroma prediction. The proposed method first introduces the channel attention module to predict the chroma components, which fully exploring the relationship between the neighboring reference sample and the predicted chroma component. Then, to make the prediction more accurate, we use multi-line reference samples. It is worth mentioning that even CACNN only uses the patches which are encoded by QP22 for training, the experiments have proved that CACNN can be easily generalized to the other unseen QPs. CACNN can be easily integrated into the VTM 8.2 as a new intra chroma prediction mode. The experimental results show that the proposed method can achieve a remarkable compression performance for Cb and Cr components in high QPs compared with VTM 8.2 anchor.

Acknowledgments. This work was supported in part by the Key Research and Development Program of China under Grant 2018YFC0831000, the National Natural Science Foundation of China under Grants 61871342, the open project program of state key laboratory of virtual reality technology and systems, Beihang University, under Grant VRLAB2021A01, and the OPPO Research Fund.

References

1. François, E., Fogg, C., He, Y., Li, X., Luthra, A., Segall, A.: High dynamic range and wide color gamut video coding in HEVC: status and potential future enhancements. IEEE Trans. Circuits Syst. Video Technol. **26**(1), 63–75 (2016)
2. Bross, B.: Versatile video coding (Draft 2). Document JVET-K1001, Ljubljana, SI, July 2018
3. Sullivan, G.J., Ohm, J., Han, W., Wiegand, T.: Overview of the high efficiency video coding (HEVC) standard. IEEE Trans. Circuits Syst. Video Technol. **22**(12), 1649–1668 (2012)
4. Li, Y., et al.: A hybrid neural network for chroma intra prediction. In: 2018 25th IEEE International Conference on Image Processing (ICIP), 2018, pp. 1797–1801(2018)
5. Blanch, M.G., Blasi, S., Smeaton, A., O'Connor, N. E., Mrak, M.: Chroma intra prediction with attention-based CNN architectures. In: 2020 IEEE International Conference on Image Processing (ICIP), 2020, pp. 783–787(2020)
6. Zhu, L., Zhang, Y., Wang, S., Kwong, S., Jin, X., Qiao, Y.: Deep learning-based chroma prediction for intra versatile video coding. IEEE Trans. Circuits Syst. Video Technol. **31**(8), 3168–3181 (2020)
7. Hu, J., Shen, L., Albanie, S., Sun, G., Wu, E.: Squeeze-and-excitation networks. IEEE Trans. Pattern Anal. Mach. Intell. **42**(8), 2011–2023(2020)
8. Nair, V., Hinton, G.E.: Rectified linear units improve restricted Boltzmann machines Vinod Nair. In: International Conference on International Conference on Machine Learning Omnipress (2010)
9. Chen, J., Ye, Y., Kim, S.: Algorithm description for versatile video coding and test model 8 (VTM 8). Document JVET-Q2002, Brussels, BE, January 2020
10. Boyce, J., Suehring, K., Li, X., Seregin, V.: JVET common test conditions and software reference configurations. Document JVET-J1010, Ljubljana, Slovenia, July 2018
11. Bjontegaard, G.: Calculation of average PSNR differences between RD-curves. ITU-T VCEG-M33, April 2001

Biometric Identification Techniques

A Novel Deep Residual Attention Network for Face Mosaic Removal

Chen Liu and Shutao Wang[✉]

Measurement Technology and Instrumentation Key Lab of Hebei Province, Yanshan University, Qinhuangdao 066004, HeBei, China
liuchen@stumail.ysu.edu.cn, wangshutao@ysu.edu.cn

Abstract. Deep learning used to achieve face mosaic removal is in full swing. In this paper, a novel deep residual attention network (DRAN) is proposed for face mosaic removal. Inspired by the application of attention mechanism, we apply channel attention (CA) and pixel attention (PA) to DRAN to make the network focus on more informative information. In addition, we improve the conventional pixel attention which we superimpose three convolutional kernels of different sizes. DRAN consists of an encoder and a decoder, which the clean and real face image is reconstructed by convolutional neural network. In the encoder, the feature maps of each convolutional layer are used as the input of CA, the output of CA is sent to PA, and the output of PA is directly concatenated with the corresponding feature maps of the decoder. As the same time, inspired by the residual learning, we propose the parallel residual block for more detailed feature extraction. Extensive experiments show that DRAN performs better than state-of-the-art methods, the best PSNR (peak signal-to-noise ratio) and SSIM (structural similarity index) based on the test set are 20.67 dB and 0.8509, respectively.

Keywords: Face mosaic removal · Image reconstruction · Residual learning · Channel attention mechanism

1 Introduction

The transformation from low-resolution (LR) face to high-resolution (HR) face has always been a hot topic in face analysis. Many researchers have made great contribution to this, and the proposed methods are worthy of reference for subsequent researches [10, 11]. With the development of image super-resolution (SR) technology, it is used in various computer vision application [12–14]. At the same time, it is also successfully applied to face mosaic removal.

Kim et al. proposed a very deep SR convolutional network (VDSR) inspired by VGGNet [26] for ImageNet classification, which can achieve accurate single-image super-resolution [1]. Affected by the depth of the network, the convergence speed will slow down [2]. To solve this problem, the authors proposed to learn residuals only and use extremely high learning rates [1]. But this method is not satisfactory when applied to face mosaic removal, especially when the degree of the blur is deep.

© Springer Nature Singapore Pte Ltd. 2021
Y. Wang and W. Song (Eds.): IGTA 2021, CCIS 1480, pp. 63–73, 2021.
https://doi.org/10.1007/978-981-16-7189-0_6

However, most image SR methods are based on faster and deeper convolutional neural networks (CNN) [1, 21], but we cannot recover the more details and it is related to the fact that most methods focus on minimizing the mean square error (MSE) [3]. Ledig et al. [3] proposed super-resolution generative adversarial network (SRGAN), and a perceptual loss function was introduced which consists of an adversarial loss and a content loss. The mean-opinion-score (MOS) is used to evaluate the quality of the images generated by SRGAN, but MOS is not objective and it is difficult to follow for fair comparison [4].

Chen et al. [4] proposed an end-to-end learning face super-resolution network (FSR-Net) with facial priors which face alignment and parsing are taken as the new evaluation metrics to address the problem of MOS in SRGAN. FSRNet successfully combines SR technology with facial priors, and it is composed of coarse SR network, fine SR encoder, prior estimation network and fine SR decoder. Two kinds of facial priors: facial land-mark heatmaps and parsing maps are introduced in this method. In order to produce more realistic HR faces, Face Super-Resolution Generative Adversarial Network (FSR-GAN) is introduced which embeds the adversarial loss into FSRNet [4]. The FSRNet achieves state-of-the-art performance, but the training set of FSRNet needs the segmentation masks of facial attributes which are based on manual annotation [24]. Obviously, it is not usually the job that one person can do, and it will be time-consuming. If the object we deal with is not a face, we still need to label other kinds of images manually.

Menon et al. [5] proposed an alternative formulation of the SR problem, and the novel SR algorithm can generate HR, realistic images which has never been in previous literatures. Self-supervised photo upsampling via latent space exploration of generative models (PULSE) is a completely self-supervised learning mode, unlike the previous works such as FSRNet which requires training on datasets of LR-HR pairs for supervised learning [5]. In previous works, some researchers optimize their methods by calculating the MSE between the generated image and the corresponding ground truth. But Menon et al. found that generated images still show signs of blurring in high variance areas of the images based on the pure MSE loss function, and this problem is improved by finding points which actually lie on the natural image manifold and downscaling correctly [5]. While LR face images need to be aligned in advance, and this process is realized through the pre-trained face key points predictor, which may produce some unexpected results [19]. For example, some LR face images cannot obtain the corresponding facial key points based on the predictor (In our 100 test face images, there are 24 face images are not detected by the predictor for any facial key points with mosaic level 5.).

In order to discard the facial priors [4, 23] and improve PULSE, we propose a novel method of face mosaic removal, that is DRAN which we propose the parallel residual block for more detailed feature extraction, and we utilize CA for treating the information of different channels differently. DRAN shows satisfactory results in terms of PSNR and SSIM, although there is still room for improvement in generating richer facial details.

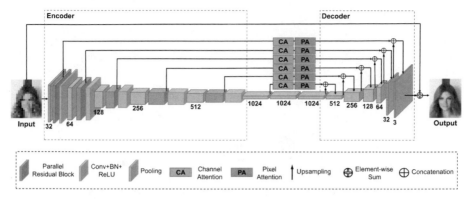

Fig. 1. The architecture of DRAN proposed in this paper.

2 The Proposed Method

In this paper, we propose a novel deep residual attention network to capture multi-scale features, and it is an end-to-end trainable network. But deeper networks are more difficult to train [2], for example, the loss may prematurely converge, and the features extracted from convolutional layers may be lost.

Inspired by the deep residual learning [2], we propose the parallel residual block to address these problems that may appear in the training process of DRAN. The face mosaic removal has always been a challenging work for researchers, in particularly, it is difficult to ensure that the removal results can restore the details of the ground truth to the greatest extent. We try to use a method different from the previous works, that is channel attention mechanism, which can make the network focus on more informative information [6], and no extra works are introduced.

What DRAN needs to do is to make the face image generated by network as close to the ground truth as possible, and we achieve it by minimizing the MSE between the generated image and the corresponding ground truth. The loss function of DRAN can be written as follows:

$$L = \frac{1}{2N} \sum_{i=1}^{N} \left\| \hat{y}_i - y_i \right\|^2 \tag{1}$$

where $\hat{y}_i$ and y_i are the face image generated by DRAN and the ground truth, respectively. N denotes the size of the batch, and the architecture of DRAN is shown in Fig. 1.

2.1 Parallel Residual Block

Residual network (ResNet) proposed by He et al. successfully solves the problem that deeper networks are more difficult to train [2]. Inspired by ResNet, we design the parallel residual block (PRB), and we applied the batch normalization (BN) to the PRB which can make the network converge quickly [7]. The architecture of the PRB is shown

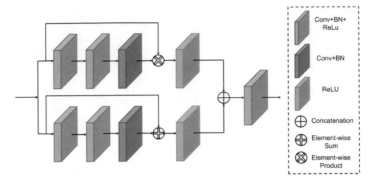

Fig. 2. The framework of parallel residual block which contains two branches.

in Fig. 2. We have changed the common residual learning, because we add a series of convolutional output to the input at the element-wise level which can treat the input information equally but make the useless information retained again. In PRB, we multiply a series of convolutional output and the input at the element-wise level, and the distinctive information is given higher weights. We can see from Fig. 3 that the element-wise product makes the contour of face, facial features and hair get more attention, and they have higher weights.

The PRB has two branches, and each branch consists of three convolutional layers, and each convolutional (Conv) layer follows BN. We select rectified linear unit (ReLU) as the activation function which is added after the Conv-BN pairs except for the last one [16]. We assume that the input of the PRB is $X = [x_1, \cdots, x_c, \cdots, x_C]$ which is equivalent to the feature maps with the size of $H \times W \times C$. The process of the PRB can be written as follows:

$$PRB_c^1 = Conv(\delta(Conv(\delta(Conv(x_c))))) \tag{2}$$

$$PRB_c^2 = Conv(\delta(Conv(\delta(Conv(x_c))))) \tag{3}$$

$$PRB_c = \delta(Conv(\delta(PRB_c^1 \cdot x_c) \oplus \delta(PRB_c^2 + x_c))) \tag{4}$$

where $\delta(\cdot)$ is the ReLU function, PRB_c^1 and PRB_c^2 are the outputs of the last Conv-BN pairs in the two branches, respectively. Then we compute the product of PRB_c^1 and x_c, and the sum of PRB_c^2 and x_c at the element-wise level, respectively. Finally, $\delta(PRB_c^1 \cdot x_c)$ is directly concatenated with $\delta(PRB_c^2 + x_c)$, and passes through the convolutional layer to get PRB_c, which is the output of PRB on the c-th channel.

2.2   Channel Attention

Previous methods based on CNN always treat the channel-wise of mosaic images equally, but this is not the case. Zhang et al. proposed a channel attention (CA) mechanism which can make the network focus on more informative information [6]. CA is widely applied to image dehazing [8] and other feature extraction works, and the architecture of CA

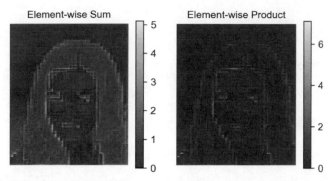

Fig. 3. Visual results of the feature maps from two branches of PRB and the corresponding weight maps.

proposed in this paper is shown in Fig. 4. The global average pooling (GAP) and global max pooling (GMP) can capture the global average common and the most characteristic information [25], respectively. They are used to take the channel-wise global spatial information into a channel descriptor at first [6], the size of input X (defined in 2.1.) is shrunk from $H \times W \times C$ to $1 \times 1 \times C$ via the GAP and GMP:

$$A_c = \mathrm{H_{GAP}}(x_c) = \frac{1}{H \times W} \sum_{i=1}^{H} \sum_{j=1}^{W} x_c(i,j) \tag{5}$$

$$M_c = \mathrm{H_{GMP}}(x_c) = \max(x_c) \tag{6}$$

where $\mathrm{H_{GAP}}(\cdot)$ and $\mathrm{H_{GMP}}(\cdot)$ denote the GAP and GMP function, respectively. $x_c(i,j)$ is the value at the position of (i,j) in the c-th feature map, A_c and M_c are the shrinking results of the c-th channel-wise feature map x_c.

Zhang et al. introduce a gating mechanism which sigmoid function is utilized due to the two criteria it meets: First, it can be able to learn nonlinear interactions between channels. Second, it can be able to learn a non-mutually-exclusive relationship [17].

$$A'_c = \sigma(Conv(\delta(Conv(A_c)))) \tag{7}$$

$$M'_c = \sigma(Conv(\delta(Conv(M_c)))) \tag{8}$$

where $\sigma(\cdot)$ and $\delta(\cdot)$ are the sigmoid function and ReLU function [16], respectively. We element-wise sum A'_c and M'_c to get CA'_c, and finally, we multiply CA'_c by x_c at an element-wise mode to get CA_c:

$$CA_c = CA'_c \cdot x_c \tag{9}$$

then $CA = [CA_1, \ldots, CA_c, \cdots, CA_C]$ can be obtained.

2.3 Pixel Attention

Pixel attention makes the network pay more attention to more important features in the pixel dimension, and it first appeared in the related work of image dehazing. Qin

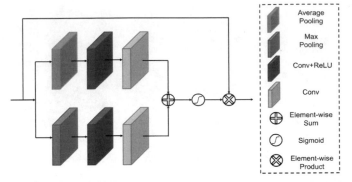

Fig. 4. The architecture of channel attention used in our paper.

et al. proposed a feature fusion attention network for single image dehazing which they applied channel attention to their work and proposed a pixel attention (PA) module. However, different sizes of convolutional kernels have different mapping regions in the input image, namely receptive field. At the pixel level, we use three convolutional kernels of different sizes to enrich the features captured by the network which is different from the conventional PA, and the architecture is shown in Fig. 5.

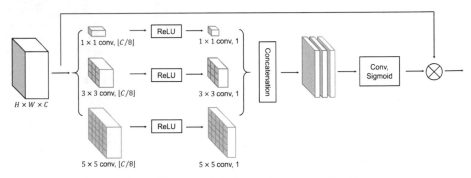

Fig. 5. The architecture of pixel attention proposed in this paper.

First, we assume that the input of PA is $H \times W \times C$, and we use convolutional kernels of 3×3, 5×5 and 7×7 to compress the channel dimension from C to $\lfloor C/8 \rfloor$ and then to 1. After that, we concatenated three different feature maps in channel dimension to get the feature maps of $H \times W \times 3$ which needs to be compressed into $H \times W \times 1$. Finally, the information of the feature maps is activated by sigmoid function, and we multiply the activated values with the input at an element-wise mode.

3   Experimental Results

3.1   Experimental Settings

Tang et al. proposed a large-scale CelebFaces attributes (CelebA) dataset which contains more than 200k celebrity images, and there are 40 attribute annotations for each. The size of each image in CelebA is 178×218, and all images cover large pose variations and background clutter [15, 20]. In the experiment, we select the first 18000 images for training, and the following 100 images for evaluation which is similar to the previous work [4]. For each face image, we add a certain level of mosaic to it, and we use α to express it. The meaning of α is shown in Fig. 6, and we set $\alpha = 5$, 10 and 15 in the experiment to compare the performance of DRAN in different mosaic levels.

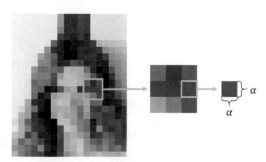

Fig. 6. The α proposed in this paper to express a level of mosaic. Each square with a side length of α contains α^2 pixels, and each pixel has the same value (In the experiment, we set $\alpha = 5$, 10 and 15, respectively.).

In the encoder, each Conv-BN-ReLU pair follows PRB, and the size of c filters is 3×3 which $C = \{3, 32, 64, 128, 256, 512, 1024\}$ and $c = \{x | x \subset C\}$, and the feature maps are down-sampled by max pooling and average pooling which the size of the filter and the size of the stride are both 2. We element-wise sum the output of max pooling and the output of average pooling, and the result is the input of PRB. In PRB, the size of c filters in Conv layers is 3×3, and the value of padding is 1 to ensure that the output size of the feature maps is the same as the input size of the feature maps. The output of PRB is the input of CA, we assume the input size of CA is $H \times W \times c$, the size of the feature maps after GAP and GMP becomes $1 \times 1 \times c$, the size of the feature maps after the first Conv layers changes from $1 \times 1 \times c$ to $1 \times 1 \times \lfloor c/8 \rfloor$, and the size of the feature maps after the second Conv layers changes from $1 \times 1 \times \lfloor c/8 \rfloor$ to $1 \times 1 \times c$. The size of all filters in CA is 1×1, and the padding is not needed.

The bilinear interpolation is used for upsampling, and the result of each upsampling is concatenated with the output of corresponding CA, and then sent to the Conv layers. The parameter settings in the decoder are the same as the encoder, and the last feature maps with 3 channels should be element-wise summed by the input mosaic image. The whole network is trained for 150, and we trained the dataset in batches, and the size of each batch is 20. We use Adam [18] optimizer which the initial learning rate is 1×10^{-4}, β_1 and β_2 take the default values of 0.9 and 0.999, respectively. While the decay strategy of learning rate has a crucial impact on training. In previous works, most researchers used linear and exponential decay strategies, at the same time, other effective decay strategies are constantly being studied. Loshchilov et al. [9] proposed a learning rate decay strategy based on cosine annealing which the learning rate changes from the initial value to 0 by following the cosine function. We assume the total number of batches is T, the initial learning rate is η, then the learning rate η_t at batch t is computed as:

$$\eta_t = \frac{1}{2}(1 + \cos(\frac{t\pi}{T}))\eta \tag{10}$$

All the experiments are carried out in the PyTorch [22] environment running on a PC with Intel Core i5-9400F CPU 2.90 GHz and a Nvidia GeForce GTX 1070Ti 8G GPU.

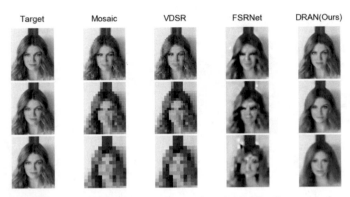

Fig. 7. Visual results of different face mosaic removal methods. The first row is the case of $\alpha = 5$, the second row is $\alpha = 10$, and the last raw is $\alpha = 15$.

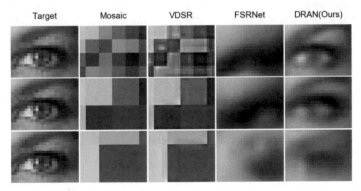

Fig. 8. Eye details of different mosaic removal methods. The first row is the case of $\alpha = 5$, the second row is $\alpha = 10$, and the last row is $\alpha = 15$.

3.2 Quantitative and Qualitative Evaluation

In this paper, we use PSNR and SSIM as the evaluation metrics to evaluate our DRAN and other state-of-the-art methods. Based on the same test face images, we compare DRAN with VDSR, FSRNet and PULSE with mosaic level 5, 10, and 15. Some visual results are shown in Fig. 7. Among the three methods in Fig.7, DRAN is obviously the best in terms of the quality of the demosaicking images. It seems that there is not much difference between VDSR and mosaic images, and VDSR is much inferior to other methods for face mosaic removal. However, the results of FSRNet are somewhat bizarre, especially the facial features and facial contour are very uncoordinated and unreal. As for PULSE, the face mosaic images need to be aligned in advance through the pre-trained face key points predictor, but the predictor may fail for some mosaic images [19]. For example, in our 100 test face images, there are 24 face images are not detected by the predictor for any facial key points when $\alpha = 5$, so PULSE can only achieve the mosaic removal of the remaining 76 face images. While when $\alpha = 10$ and 15, the predictor fails for all 100 mosaic images, and this may be a drawback of PULSE. From Fig. 8, we can see that DRAN is the best in detail processing of the face compared with the other two methods, and no other noise is introduced. We synthesize the average PSNR and SSIM of VDSR, FSRNet and DRAN on the test set with different mosaic levels, and show them in Table 1. It is obviously that our DRAN is the best in both PSNR and SSIM in Table 1. Especially SSIM, DRAN is far superior to other methods and the SSIM of DRAN keeps above 0.7000 at three different mosaic levels.

Table 1. Average PSNR (dB)/SSIM results of different methods for face mosaic removal with mosaic level (α) 5, 10 and 15.

Mosaic level	VDSR	FSRNet	DRAN
	PSNR/SSIM	PSNR/SSIM	PSNR/SSIM
5	19.49/0.5565	17.81/0.5495	20.67/0.8509
10	16.68/0.4794	15.93/0.4255	16.77/0.7585
15	14.59/0.4554	14.42/0.3437	15.07/0.7101

4   Conclusion

The proposed DRAN is different from the previous image demosaicking methods, and we have made a lot of improvements to the previous methods, for example, we discard the generative adversarial network which is easy to produce blurry images, and no facial priors are needed. For the latest method PULSE, it depends on the face key points predictor, and we prove that the predictor will fail for some or even all face mosaic images, but DRAN can achieve the face mosaic removal for all images with different mosaic levels. In terms of the quality of the demosaicking images, DRAN is acceptable and satisfactory, but we still need to focus on how to maximize the detailed texture of the face images, including hair, facial features, beard, earrings and so on.

References

1. Kim, J., Lee, J.K., Lee, K.M.: Accurate image super-resolution using very deep convolutional networks. In: IEEE Conference on Computer Vision & Pattern Recognition, pp. 1646–1654 (2016)

2. He, K., Zhang, X.Y., Ren, S.Q., et al.: Deep residual learning for image recognition. In: IEEE Conference on Computer Vision & Pattern Recognition (CVPR), pp. 770–778 (2016)
3. Ledig, C., Theis, L., Huszr, F., et al.: Photo-realistic single image super-resolution using a generative adversarial network. In: IEEE Conference on Computer Vision & Pattern Recognition (CVPR), pp. 105–114 (2017)
4. Chen, Y., Tai, Y., Liu, X.M., et al.: FSRNet: end-to-end learning face super-resolution with facial priors. In: IEEE Conference on Computer Vision & Pattern Recognition (CVPR), pp. 2492–2501 (2018)
5. Menon, S., Damian, A., Hu, S., et al.: PULSE: self-supervised photo upsampling via latent space exploration of generative models. In: IEEE Conference on Computer Vision & Pattern Recognition (CVPR), pp. 2434–2442 (2020)
6. Zhang, Y., Li, K., Li, K., Wang, L., Zhong, B., Fu, Y.: Image super-resolution using very deep residual channel attention networks. In: Ferrari, V., Hebert, M., Sminchisescu, C., Weiss, Y. (eds.) ECCV 2018. LNCS, vol. 11211, pp. 294–310. Springer, Cham (2018). https://doi.org/10.1007/978-3-030-01234-2_18
7. Ioffe, S., Szegedy, C.: Batch normalization: Accelerating deep network training by reducing internal covariate shift (February 2015)
8. Qin, X., Wang, Z.L., Bai, Y.C., et al.: FFA-Net: feature fusion attention network for single image dehazing. In: Proceedings of the AAAI Conference on Artificial Intelligence, vol. 43, no. 7 (2020)
9. Loshchilov, H., Hutter, F.: SGDR: stochastic gradient descent with warm results. In: International Conference on Learning Representations (2017)
10. Nasrollahi, K., Moeslund, T.B.: Super-resolution: a comprehensive survey. Mach. Vis. Appl. 25(6), 1423–1468 (2014)
11. Yang, C.-Y., Ma, C., Yang, M.-H.: Single-image super-resolution: a benchmark. In: Fleet, D., Pajdla, T., Schiele, B., Tuytelaars, T. (eds.) ECCV 2014. LNCS, vol. 8692, pp. 372–386. Springer, Cham (2014). https://doi.org/10.1007/978-3-319-10593-2_25
12. Sajjadi, M.S., Schlkopf, B., Hirsch, M.: Enhancement: single image super-resolution through automated texture synthesis. In: IEEE International Conference on Computer Vision (ICCV) (2016)
13. Shi, W., et al.: Cardiac image super-resolution with global correspondence using multi-atlas patchmatch. In: Mori, K., Sakuma, I., Sato, Y., Barillot, C., Navab, N. (eds.) MICCAI 2013. LNCS, vol. 8151, pp. 9–16. Springer, Heidelberg (2013). https://doi.org/10.1007/978-3-642-40760-4_2
14. Zou, W.W., Yuen, P.C.: Very low-resolution face recognition problem. IEEE Trans. Image Process. 21(1), 327–340 (2012)
15. Liu, Z., Luo, P., Wang, X., Tang, X.: Deep learning face attributes in the wild. In: IEEE International Conference on Computer Vision (ICCV), pp. 3730–3738 (2015)
16. Nair, V., Hinton, G.E.: Rectified linear units improve restricted Boltzmann machines. Int. Conf. Mach. Learn. 27, 807–814 (2012)
17. Hu, J., Shen, L., Sun, G.: Squeeze-and-excitation networks. IEEE Trans. Pattern Anal. Mach. Intell. 42(8), 2011–2023 (2019)
18. Kingma, D., Ba, J.: Adam: a method for stochastic optimization. In: International Conference on Learning Representations (2014)
19. Zoeller, G., Damian, A., Clauss, C., et al.: PULSE: self-supervised photo upsampling via latent space exploration of generative methods (2020). https://github.com/adamian98/pulse
20. Liu, Z., Luo, P., Wang, X., Tang, X.O.: Large-scale celebfaces attributes (CelebA) dataset. http://mmlab.ie.cuhk.edu.hk/projects/CelebA.html
21. Dong, C., Loy, C.C., He, K., Tang, X.: Learning a deep convolutional network for image super-resolution. In: Fleet, D., Pajdla, T., Schiele, B., Tuytelaars, T. (eds.) ECCV 2014. LNCS, vol. 8692, pp. 184–199. Springer, Cham (2014). https://doi.org/10.1007/978-3-319-10593-2_13

22. Paszke, A., Gross, S., Massa, F., Lerer, A., Chintala, S.: PyTorch: an imperative style, high-performance deep learning library. Adv. Neural Inf. Process. Syst. **32**, 8026–8037 (2019)
23. Kim, D., Kim, M., Kwon, G., Kim, D.S.: Progressive face super-resolution via attention to facial landmark. In: Proceedings of the 30th British Machine Vision Conference (BMVC) (2019)
24. Liu, Z., Lee, C.: CelebAMask-HQ (2020). https://github.com/switchablenorms/CelebAMask-HQ
25. Li, P., Tian, J., Tang, Y., et al.: Deep retinex network for single image dehazing. IEEE Trans. Image Process. **30**, 1100–1115 (2020)
26. Simonyan, K., Zisserman, A.: Very deep convolutional networks for large-scale image recognition, Computer Science (2014)

Machine Vision and 3D Reconstruction

Pretrained Self-supervised Material Reflectance Estimation Based on a Differentiable Image-Based Renderer

Tianteng Bi[1], Yue Liu[1,2(✉)], Dongdong Weng[1,2], and Yongtian Wang[1,2]

[1] Beijing Engineering Research Center of Mixed Reality and Advanced Display,
School of Optics and Photonics, Beijing Institute of Technology,
Beijing 100081, China
{bitianteng,liuyue,crgj,wyt}@bit.edu.cn
[2] AICFVE of Beijing Film Academy,
4 Xitucheng Road, Haidian, Beijing 100088, China

Abstract. Measuring the material reflectance of surfaces is a key technology in inverse rendering, which can be used in object appearance reconstruction. In this paper we propose a novel deep learning-based method to extract material information represented by a physically-based bidirectional reflectance distribution function from an RGB image of an object. Firstly, we design new deep convolutional neural network architectures to regress material parameters by self-supervised training based on a differentiable image-based renderer. Then we generate a synthetic dataset to train the model as the initialization of the self-supervised system. To transfer the domain from the synthetic data to the real image, we introduce a test-time training strategy to finetune the pretrained model to improve the performance. The proposed architecture only requires one image as input and the experiments are conducted to evaluate the proposed method on both the synthetic data and real data. The results show that our trained model presents dramatic improvement and verifies the effectiveness of the proposed methods.

Keywords: Material prediction · Inverse rendering · Deep learning

1 Introduction

Acquiring the material reflectance of an object is very important for such applications as mixed reality, robotics, and artistic creation. The reflectance of surface material requires complicated optical devices to conduct dense measurements of the target object [1]. However, such devices can usually only be used for a certain class of objects and the measurements need an extremely strict experiment environment, which is a costly effort and merely used in a few scenes.

A more general and efficient method with the name of inverse rendering is to infer the material from the objects' images [2]. However, this is a highly ill-posed problem as quite a few combinations among the ambient light, geometry, and

© Springer Nature Singapore Pte Ltd. 2021
Y. Wang and W. Song (Eds.): IGTA 2021, CCIS 1480, pp. 77–91, 2021.
https://doi.org/10.1007/978-981-16-7189-0_7

material may lead to the same observed image. How to address the ambiguities is the key to acquire these properties from images. Benefiting from the development of deep learning, many methods have been proposed to estimate the material reflectance using a single image. However, training such deep models requires large amounts of labeled data that are extremely difficult to acquire in practice. Most of the existing methods usually adopt computer-generated data to train the model instead of the real captured one, but the models trained by the synthetic data suffer from the domain gap between the different datasets, which presents poor performance over the captured dataset.

To tackle the problem of the absence of real labeled data, this paper extracts material information from an object's image using deep convolutional neural networks (CNN) by self-supervised learning, which is initialized with the pretrained model trained by the synthetic data. The self-supervised architecture based on a differentiable image-based renderer is designed and the synthetic dataset used for initializing the CNN is generated. To overcome the domain gap with good performance, a test-time training strategy is also introduced to our designed architecture.

In summary, we make the following two main contributions:

- We design two self-supervised material capturing architectures with a physically based differentiable renderer, which can be implemented with variable CNN. The performance of the designed models is evaluated with the synthetic data and the real one.
- We propose an approach of initializing the self-supervised model to solve the ambiguity introduced by the random initialization. Combining with the test-time training strategy, the performance of the model over real data is improved by a domain adaption procedure.

2   Related Work

Inverse rendering is a fundamental problem in computer vision and graphics, which is a highly ill-posed problem due to the mixture of material, geometry, and ambient light. Among these properties, the material is more complicated to represent and describe. According to the interaction process of light and matter, the appearance of objects with isotropic and opaque surfaces is usually described by bidirectional reflection distribution function (BRDF), which physically represents the ratio of the reflected energy from a point on the surface to the incident energy. Some existing methods adopt the automatic optical equipment to conduct dense measurements of BRDF with an object. Marschner et al. construct a hand-held camera measurement system mainly equipped with the CCD sensor, RGB color filter array, and industrial electronic flash to measure simple geometric objects such as sphere and cylinder [3]. The gonioreflectometer is also used for measuring BRDF, which is composed of the light source, material sample rack, turntable, and detector [4,5]. Catadioptric systems adopting reflected light and refracted light to eliminate aberration and spherical or hemispherical gantry structures are also suitable for dense BRDF measurement [6–8]. The method of using such acquisition equipment for measurement needs a strict experimental environment and the efficiency is low, thus it is only suitable for a few objects with simple material and shape, and the practical application scene is very limited.

Image-based BRDF estimation methods utilize one or few images to obtain the reflectance, but it suffers from the shortcomings of ambiguities. Some existing methods solve this challenging problem by constructing a set of priors over the constituted properties or assuming one of the object's geometry and the lighting condition being known, and then iteratively optimize a hand-crafted mathematical model to find the solutions [9–11]. Compared with the above-mentioned methods, recently CNN makes great progress in this task. A deep lambertian model is proposed by Tang et al. to predict diffuse material, point light direction, and orientation map from a single image with Gaussian Restricted Boltzmann Machines [12]. Georgoulis et al. predict the normal maps and reflectance maps from a single image with the designed CNN [13]. Liu et al. design three separate CNNs to predict the material parameters, normal maps as well as environment maps from a single image, which is followed by a differentiable rendering layer [14]. However, these methods use synthetic data to train their CNN, which makes the model suffers from a domain gap leading to a low performance over real data.

3   Self-supervised Architectures

Training the CNN model requires a lot of annotation data to obtain material reflection properties, but in practice, measuring the reflectance of object surface material is extremely complex and tedious, and even impossible in some cases.

To solve such a problem, this paper uses an embedded image-based renderer to construct two self-supervised architectures to obtain the reflectance of surface material from an object's 2D image without real labeled data. Some existing works verify that the differentiable renderer can be embedded in the CNN to construct the perceptual reconstruction loss between the input and the output, which works as an additional constraint to generate better results in the view of perception for inverse rendering [15, 16].

Assuming that the surface of an object is composed of a single material and the reflectance of the material is represented by a parametric BRDF, obtaining the reflection properties from the image can be regarded as mapping the 2D image of the object to the BRDF parameters. Specifically, the designed self-supervised architectures for material reflectance obtaining are composed of an encoder and an image-based renderer as shown in Fig. 1. The encoder maps the input image to a vector that represents the material and the renderer takes the vector as input to render an input-like image given the ambient light and geometry information.

The first self-supervised architecture is composed of a single encoder and an image-based renderer. The encoder outputs the material parameters and the renderer renders an image as shown in Fig. 1(a). The second self-supervised architecture is the combination of the Siamese network and two renderers. In this architecture, the input is a pair of images and the Siamese encoder maps the two inputs to two vectors. The rendering is implemented by the dual renderer which outputs two rendered images. The Siamese network is composed of two CNN with the same structure and has the same weights through the weights sharing mechanism. It was originally used to measure the similarity degree of the two inputs [17].

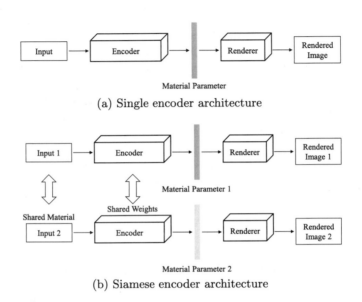

(a) Single encoder architecture

(b) Siamese encoder architecture

Fig. 1. Diagram of two self-supervised architecture

4   Image-Based Differentiable Renderer

In the self-supervised training architecture, the core of the system is the image-based renderer which can be embedded into CNN to generate images. In this paper, the physical image formation procedure is adopted to construct the renderer [18], which can be formulated as the rendering equation:

$$L_o(x, \vec{\omega}_o) = L_e(x, \vec{\omega}_o) + L_r(x, \vec{\omega}_o)$$
$$= L_e(x, \vec{\omega}_o) + \int_{\Omega+} f(x, \vec{\omega}_i, \vec{\omega}_o) L_i(x, \vec{\omega}_i)(\vec{\omega}_i \cdot \vec{n}_x) d\vec{\omega}_i \qquad (1)$$

where L_o expresses the radiance leaving the point x with normal $\vec{n}_x$ in direction $\vec{\omega}_o$ as the emitted L_e and reflected radiance L_r, which is a function of incoming light L_i over the hemisphere from direction $\vec{\omega}_i$. f represents the BRDF which describes the material in this paper. For objects that do not emit light themselves, all the emitted light is reflected, namely $L_e(x, \vec{\omega}_o) = 0$.

4.1   Normal Representation

We adopt the normal map to store the normal information of the object. Specifically, the normal map is a $H \times W \times 3$ map of which each channel stores the x, y, and z coordinate respectively of the point on the object for an image of $H \times W \times 3$.

4.2   Ambient Light Representation

The panoramic high dynamic range (HDR) environment map is used to represent the fixed incident light [19]. The environment map is based on the assumption that the light source is infinite and there is no light emitted or reflected from the object into the environment, and usually does not consider the internal reflection of the object.

As shown in Fig. 2 the spherical panorama can be flattened into a 2D image. Each pixel in the image can be mapped to the spherical coordinate and the corresponding incident direction can be calculated. Each pixel value represents the light intensity in the incident direction, and the Z-axis of the coordinate in the image points to the screen direction.

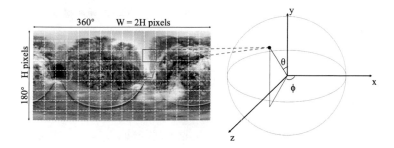

Fig. 2. Panaramic environment map

Assuming that H and W represent the height and width of the environment map respectively, for the pixel i in the environment map, h and w represent the index of the pixel, and its coordinates in the spherical coordinate are:

$$\begin{cases} \theta_i = \dfrac{h}{H}\pi \\ \phi_i = \dfrac{w}{W}2\pi \end{cases} \tag{2}$$

In addition, the weight dw_i of the pixel i can be computed as the ratio of latitude perimeter to equatorial perimeter:

$$dw_i = \frac{2\pi R \sin\theta_i}{2\pi R} = \sin\theta_i \tag{3}$$

4.3   Material Representation

In this paper, we focus on opaque objects without considering transmitted and scattered light, so the material is represented by the reflectance that can be fully formulated as BRDF that describes how the incident light is reflected off the surface. It can be defined as:

$$f(x, \vec{\omega}_i, \vec{\omega}_o) = \frac{dL(x, \vec{\omega}_o)}{dE(x, \vec{\omega}_i)} \qquad (4)$$

which is the ratio of the radiance dL leaving the surface at point x in direction $\vec{\omega}_o$ and irradiance dE arriving at x from direction $\vec{\omega}_i$.

For all kinds of opaque materials in the real world, this paper uses the directional statistics bidirectional reflection distribution function (DSBRDF) model to describe the reflectance, which has a small set of parameters and an analytic expression to model a wide range of real-world isotropic BRDF accurately [20]. Compared with non-parametric models adopting a lookup table to store the reflectance information and the existing micro-facets-based models, DSBRDF achieves higher accuracy without the linear combination of different parametric models and can be differentiated to support the back-propagation in deep learning. Specifically, the DSBRDF model is composed of a set of hemispherical exponential power distributions known as lobes that enable encoding a variety of the BRDFs. In a typical setting, the number of the lobes is 3 and there are 108 coefficients in total in such a model.

Based on the above-mentioned elements representation, DSBRDF based on renderer can be formulated as:

$$I_p^c(\vec{L}^c, \vec{n}_p, \vec{m}_\kappa^{l,c}, \vec{m}_\gamma^{l,c}) = \sum_{i=1}^{N} (\sum_{l=0}^{3} (e^{\kappa_l(\theta_d; \vec{m}_\kappa^{l,c})(\vec{h} \cdot \vec{n}_p)^{\gamma_l^c(\theta_d; \vec{m}_\gamma^{l,c})}} - 1)) L_i^c(\vec{\omega}_i \cdot \vec{n}_p) dw_i \quad (5)$$

where I_p^c is the pixel value of the point p on the object, $c = 0, 1, 2$ represents the RGB channel, $\vec{n}_p$ is the normal of point p, $\vec{m}_\kappa^{l,c}$ and $\vec{m}_\gamma^{l,c}$ are the parameters of DSBRDF, and $\vec{L}$ is the vector form of the environment map.

The back-propagation can be implemented by computing the derivatives of the rendering equation.

5   Synthetic Dataset

CNNs require a lot of labeled data to train and the collecting of real material data is often very complicated. To solve the problem of insufficient training data, computer-generated data is usually used for training CNN, because they can be labeled according to the requirement of the task. In this paper, we generate a synthetic dataset for evaluating the performance of our self-supervised architectures and verifying the improvement by our pretrained initialization.

The dataset is generated with the DSBRDF based rendering equation (Eq. 5). Rendering an image requires normal maps, material parameters, and environment maps. In this paper, we use the public normal dataset and divide it into

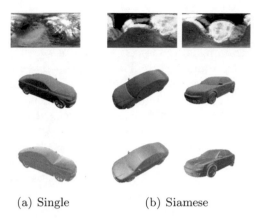

(a) Single   (b) Siamese

Fig. 3. The example of our synthetic dataset

training data and test data according to the defined classification [13]. The resolution of the normal image in the dataset is $256 \times 256 \times 3$. Besides, the normal map is normalized to $[-1, 1]$. In order to simulate the illumination of the real environment, 100 real outdoor environment maps are downloaded from the Internet. We define a split to classify the environment maps into 80 training samples and 20 testing samples according to their scenes. Moreover, random rotation is adopted to augment the lighting data. To balance the rendering quality and efficiency, the resolution is scaled from $512 \times 1024 \times 3$ to $64 \times 128 \times 3$. For the material, we use the DSBRDF parameters which are obtained by fitting to the MERL dataset [1].

A total of 50,440 training images and 9,930 testing images is generated and no materials, normal maps, and environment maps are shared between them. In addition, for the self-supervised networks, we generate the special data according to the desired input. The visualized examples from our synthetic dataset are shown in Fig. 3.

Besides, verifying the self-supervised framework on the generated dataset, we also use the real captured image to evaluate our methods [13]. An inverse tone mapping method based on deep learning is also used to preprocess the images in the dataset and transform them into HDR images [21] (Fig. 4).

(a) Captured   (b) HDR

Fig. 4. Captured image and its HDR version

6   Loss Functions

The self-supervised training is supervised by the Euclidean distance between the rendered image and the input image. The rendered image can be generated by the rendering equation with the HDR environment map, therefore the value of each pixel in the rendered image represents the intensity of the reflected light by the material, which is also HDR images. However, the numerical range and physical meaning of pixels in the low dynamic range (LDR) input image are completely different from that in HDR rendering image, and there is a nonlinear process of dynamic range mapping between them. Taking the difference into consideration, two loss functions are defined. The first one is based on HDR input:

$$\mathcal{L}_{rec}^{HDR} = \|\ln(1 + I_{rendering}) - \ln(1 + I_{input})\|_2 \tag{6}$$

where $I_{rendering}$ and I_{input} represent the rendered image and input image respectively. We also map the dynamic range in the logarithmic domain to avoid the error introduced by the excessively high dynamic range.

The second is the HDR mapping loss function for LDR input image, which linearly scale the HDR rendered image for different color channels so that the pixel value is consistent with the value range of the LDR image in the range of $[0, 255]$:

$$\mathcal{L}_{rec}^{LDR} = \|\lambda I_{rendering} - I_{input}\|_2 \tag{7}$$

where λ is the channel-wise normalization factor. It is defined as:

$$\lambda^k = \frac{255 * I_{rendering}^k}{\max I_{rendering}^k} \tag{8}$$

where $k = 0, 1, 2$ is RGB channels.

When training the single encoder architecture, Eq. (6) or Eq. (7) is used as the loss function. When training the Siamese encoder architecture, the Siamese encoder maps two input images into two material parameter vectors, which can be used to render two images. When the two input images are rendered separately, the trend of optimization is inconsistent, which will stop the convergence of CNN. Consequently, one of the input images is used as the benchmark, and two material parameters are used to render such an input image with the same normal and HDR environment map. In addition to the loss function of image reconstruction, the two predicted material parameters of the Siamese network $\vec{m}_1'$ and $\vec{m}_2'$ can be used to construct the loss of view and illumination invariance:

$$\mathcal{L}_{material}^{Siamese} = \left\| \vec{m}_1' - \vec{m}_2' \right\|_2 \tag{9}$$

Finally, the joint loss function is used to train the Siamese encoder architecture, and the HDR joint loss function is:

$$\mathcal{L}_{Siamese} = \sigma_1 \mathcal{L}_{material}^{Siamese} + \sigma_2 \mathcal{L}_{rec}^{HDR}(\vec{m}_1') + \sigma_3 \mathcal{L}_{rec}^{HDR}(\vec{m}_2') \tag{10}$$

The LDR joint loss function is:

$$\mathcal{L}_{Siamese} = \sigma_1 \mathcal{L}_{material}^{Siamese} + \sigma_2 \mathcal{L}_{rec}^{LDR}(\vec{m}_1') + \sigma_3 \mathcal{L}_{rec}^{LDR}(\vec{m}_2') \qquad (11)$$

where $\mathcal{L}_{rec}(\vec{m}_1')$ and $\mathcal{L}_{rec}(\vec{m}_2')$ represent the image reconstruction loss function with the material parameters $\vec{m}_1'$ and $\vec{m}_2'$ as well as the same normal and environment map, which are defined as Eq. (6) and Eq. (7).

7 Experiments

The experiments are conducted in Caffe framework [22], using adadelta solver on GTX2080Ti graphics card. The initial learning rate is set to 0.1 and decreases with the increase of iterations. All networks are initialized randomly.

In order to eliminate the influence of network depth, the encoder part of the two architectures refers to the network of Liu's work [14]. However, the implementation of the self-supervised architecture is not unique and the other network can also be suitable for the architecture.

Existing research results show that a self-supervised task can achieve performance improvement through test time training strategy [23,24]. In order to make the above-mentioned self-supervised architectures achieve better performance, the single target test time training strategy is introduced in the experiment.

For the synthetic dataset, the quantitative results are given by the mean square error (MSE) between predicted parameters and the groundtruth as well as MSE and the Structural Similarity(SSIM) [25] between rendered images with the predicted material and the input. In addition, the encoder in the first self-supervised architecture is trained separately in a supervised way to show the difference between self-supervised learning and supervised learning. The loss function of the encoder is the Euclidean distance between the predicted material parameter vector and the groundtruth.

Since the real captured image does not have the groundtruth of the material, we only show the quantitative results through MSE and SSIM between the rendered image and input image. The normal map and HDR environment maps required by the rendering are obtained by the U-Net that is trained with our synthetic dataset [26].

Table 1 shows the results of supervised and two self-supervised architectures over the synthetic dataset. It can be seen from the result in Table 1 that the accuracy of the self-supervised framework is lower than that of supervised learning. In self-supervised learning, the result of the LDR input image is worse than that of the HDR input image, which indicates that it is more difficult to obtain material information from LDR image only depending on image reconstruction loss. The mapping between dynamic ranges is mixed in the training, thus increasing the complexity of the task. The self-supervised learning takes the HDR rendered image as the output, which is more sensitive to the dynamic range of the input image. The HDR input image is more helpful to convergence and optimization. In contrast, supervised learning is not sensitive to dynamic range. Besides, the

Table 1. The results of the proposed self-supervised architectures on the synthetic dataset

	Material		Rendered image			
	MSE		SSIM		MSE	
	HDR	LDR	HDR	LDR	HDR	LDR
Supervised	**4.2359**	**6.5146**	**0.9703**	**0.9718**	**169.2571**	**148.5959**
Single encoder self-supervised	13.5179	19.6680	0.8835	0.8243	2153.8299	2606.65
Siamese encoder self-supervised	12.8039	17.5599	0.8827	0.8364	2204.4459	3237.17

Siamese architecture cannot provide obvious advantages but has a more complicated network than the single encoder architecture.

The qualitative results of the three architectures on the synthetic dataset and real dataset under different dynamic range input are shown in Fig. 5, from left to right are (a) input image, (b) supervised learning with HDR input (SE. H), (c) supervised learning with LDR input (SE. L), (d) single encoder self-supervised architecture with HDR input (SES. H), (e) single encoder self-supervised architecture with LDR input (SES. L), (f) Siamese encoder self-supervised architecture with HDR input (SiES.H) and (g) Siamese encoder self-supervised architecture with LDR input (SiES.L). It can also be seen from the result that the rendering result is quite different from the input image. Although the loss function converges during the training, the obtained material is not consistent with the image. The possible reason is that the random initialization at the beginning of training leads to the uncertainty of the subsequent optimization, which means that the randomly initialized CNN may not be able to find the corresponding material parameters only with the image reconstruction loss.

Table 2 shows the quantitative results on the real dataset. For supervised architecture, real images are tested directly on a trained model with our synthetic data. It can be seen from Table 2 that the result of HDR input is better than that of LDR input, which verifies that CNN combined with embedded image-based renderer is more sensitive to dynamic range, and HDR input is more helpful to the optimization. Besides, it can be found from the results of the real image in Fig. 5 that the error of supervised learning is larger than that of self-supervised architecture in the case of HDR input. The reason is that the distribution of the real image dataset and the synthetic dataset is different, resulting in the data domain mismatch on the supervised model. Since the self-supervised architectures are not trained through the synthetic datasets, they can learn the real data distribution.

It can be seen from the experimental results that the result of supervised learning is better than that of single target self-supervised learning, but the result of single target self-supervised learning is better when it is used for real images. It can be inferred that the combination of the above two processes can

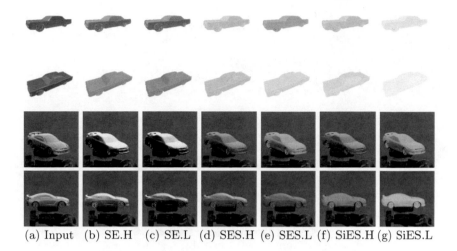

(a) Input   (b) SE.H   (c) SE.L   (d) SES.H   (e) SES.L   (f) SiES.H (g) SiES.L

Fig. 5. The comparison between different architectures on synthetic and real dataset

Table 2. The results of the proposed self-supervised architectures on the real dataset

	Rendered image			
	SSIM		MSE	
	HDR	LDR	IIDR	LDR
Supervised	0.8741	0.8674	555.2043	**625.4508**
Single encoder self-supervised	**0.8928**	**0.8730**	**368.1826**	1412.0228
Siamese encoder self-supervised	0.8847	0.8689	484.2412	1834.7536

obtain the reflectance of materials more accurately in the real image. Firstly, the correct optimization direction is determined through the supervised learning of synthetic data, which avoids the uncertainty caused by random initialization, and then the data domain is transferred to the real image based on the single target self-supervised learning.

Based on such assumption, we firstly use the supervised training model with HDR input to initialize the single encoder self-supervised architecture and then perform the single target test-time training strategy. The results can be found in Table 3 and Fig. 6. The qualitative results of the supervised and pretrained self-supervised model on the real dataset are shown in Fig. 6, from left to right are (a) input image, (b) supervised learning with HDR input (SE.H), (c) pretrained self-supervised learning with HDR input (PSES.H).

It can be seen from Table 3 and Fig. 6 that the obtained material by the final results are better, and the material, as well as specular highlight, are consistent with the original image. Compared with the random initialization, the self-supervised architecture based on pretrained initialization can learn the results

Table 3. The result of pretrained initialization

	Rendered image	
	SSIM	MSE
Supervised	0.8741	555.2043
Self-supervised	**0.8959**	**344.9891**

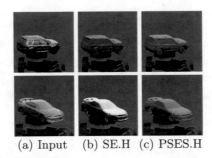

(a) Input   (b) SE.H   (c) PSES.H

Fig. 6. The comparison of supervised model and pretrained self-supervised model

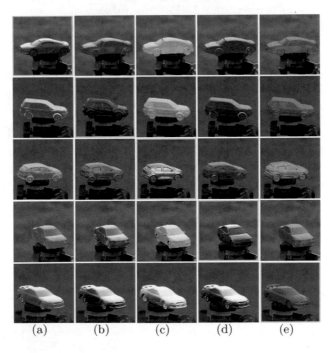

(a)   (b)   (c)   (d)   (e)

Fig. 7. Image editing with our proposed method (Color figure online)

correctly. More results can be seen in Fig. 7 which shows the cases of editing the material using our proposed method. The image marked with a yellow box in the diagonal direction is the real captured image, and the rest are rendered images.

8   Conclusion

To solve the problem of insufficient material annotation data, an embedded image renderer based on DSBRDF is developed and two self-supervised architectures are proposed. Based on the experiments over the synthetic and real dataset, uncertainty caused by random initialization in self-supervised learning and the domain gap between the synthetic data and real data in supervised learning are revealed. By initializing the single encoder self-supervised architecture with the pretrained model using synthetic data, we improve the self-supervised performance on real data combining the single target test-time training, which is verified by the quantitative and qualitative results.

Currently, DSBRDF is still a hand-crafted model and its physical meaning is not clear enough, the types of materials that can be expressed by DSBRDF are also limited. In the future, the appearance description model based on deep learning should be studied to optimize the ability for representing the material.

Acknowledgment. This work was supported by the Key-Area Research and Development Program of Guangdong Province (No. 2019B010149001) and the National Natural Science Foundation of China (No. 61960206007) and the 111 Project (B18005).

References

1. Matusik, W., Pfister, H., Brand, M., McMillan, L.: A data-driven reflectance model. ACM Trans. Graph. **22**(3), 759–769 (2003). https://doi.org/10.1145/882262.882343
2. Khan, E.A., Reinhard, E., Fleming, R.W., Bülthoff, H.H.: Image-based material editing. ACM Trans. Graph. **25**(3), 654–663 (2006). https://doi.org/10.1145/1141911.1141937
3. Marschner, S.R., Westin, S.H., Lafortune, E.P.F., Torrance, K.E.: Image-based bidirectional reflectance distribution function measurement. Appl. Opt. **39**(16), 2592–2600 (2000). https://doi.org/10.1364/AO.39.002592
4. Hsia, J.J., Richmond, J.C.: Bidirectional reflectometry. Part I: A high resolution laser bidirectional reflectometer with results on several optical coatings. J. Res. Natl. Bureau Standards. Section A, Phys. Chem. **80**(2), 189 (1976)
5. Li, H., Foo, S.C., Torrance, K.E., Westin, S.H.: Automated three-axis gonioreflectometer for computer graphics applications. In: Duparre, A., Singh, B., Gu, Z.H. (eds.) Proceedings of the Advanced Characterization Techniques for Optics, Semiconductors, and Nanotechnologies II, vol. 5878, pp. 221–231. International Society for Optics and Photonics, SPIE (2005). https://doi.org/10.1117/12.617589
6. Dana, K.J., van Ginneken, B., Nayar, S.K., Koenderink, J.J.: Reflectance and texture of real-world surfaces. ACM Trans. Graph. **18**(1), 1–34 (1999). https://doi.org/10.1145/300776.300778
7. Ghosh, A., Achutha, S., Heidrich, W., O'Toole, M.: BRDF acquisition with basis illumination. In: Proceedings of the 2007 IEEE International Conference on Computer Vision, pp. 1–8. IEEE. Piscataway (2007). https://doi.org/10.1109/ICCV.2007.4408935

8. Malzbender, T., Gelb, D., Wolters, H.: Polynomial texture maps. In: Proceedings of the 28th Annual Conference on Computer Graphics and Interactive Techniques, SIGGRAPH 2001, pp. 519–528. Association for Computing Machinery, New York (2001). https://doi.org/10.1145/383259.383320

9. Barron, J.T., Malik, J.: Shape, illumination, and reflectance from shading. IEEE Trans. Pattern Anal. Mach. Intell. **37**(8), 1670–1687 (2015)

10. Lombardi, S., Nishino, K.: Reflectance and Natural Illumination from a Single Image. In: Fitzgibbon, A., Lazebnik, S., Perona, P., Sato, Y., Schmid, C. (eds.) ECCV 2012. LNCS, vol. 7577, pp. 582–595. Springer, Heidelberg (2012). https://doi.org/10.1007/978-3-642-33783-3_42

11. Lombardi, S., Nishino, K.: Single image multimaterial estimation. In: Proceedings of the 2012 IEEE Conference on Computer Vision and Pattern Recognition, pp. 238–245. IEEE. Piscataway (2012)

12. Tang, Y., Salakhutdinov, R., Hinton, G.: Deep Lambertian networks. In: Proceedings of the 29th International Conference on International Conference on Machine Learning, ICML 2012, Madison, WI, USA, pp. 1419–1426. Omnipress (2012)

13. Georgoulis, S., et al.: Reflectance and natural illumination from single-material specular objects using deep learning. IEEE Trans. Pattern Anal. Mach. Intell. **40**(8), 1932–1947 (2018). https://doi.org/10.1109/TPAMI.2017.2742999

14. Liu, G., Ceylan, D., Yumer, E., Yang, J., Lien, J.M.: Material editing using a physically based rendering network. In: Proceedings of the 2017 IEEE Conference on International Conference on Computer Vision. IEEE, Piscataway (2017)

15. Meka, A., et al.: Lime: live intrinsic material estimation. In: Proceedings of the 2018 IEEE/CVF Conference on Computer Vision and Pattern Recognition, pp. 6315–6324. IEEE. Piscataway (2018)

16. Deschaintre, V., Aittala, M., Durand, F., Drettakis, G., Bousseau, A.: Single-image SVBRDF capture with a rendering-aware deep network. ACM Trans. Graph. (SIGGRAPH Conference Proceedings) **37**(128), 15 (2018)

17. Bromley, J., et al.: Signature verification using a "Siamese" time delay neural network. Int. J. Pattern Recogn. Artif. Intell. **07**(04), 669–688 (1993). https://doi.org/10.1142/S0218001493000339

18. Kajiya, J.T.: The rendering equation. In: Proceedings of the 13th Annual Conference on Computer Graphics and Interactive Techniques, SIGGRAPH 1986, pp. 143–150. Association for Computing Machinery, New York (1986). https://doi.org/10.1145/15922.15902

19. Debevec, P.: Rendering synthetic objects into real scenes: bridging traditional and image-based graphics with global illumination and high dynamic range photography. In: Proceedings of the 25th Annual Conference on Computer Graphics and Interactive Techniques, SIGGRAPH 1998, pp. 189–198. Association for Computing Machinery, New York (1998). https://doi.org/10.1145/280814.280864

20. Nishino, K., Lombardi, S.: Directional statistics-based reflectance model for isotropic bidirectional reflectance distribution functions. J. Opt. Soc. Am. A **28**(1), 8–18 (2011). https://doi.org/10.1364/JOSAA.28.000008

21. Eilertsen, G., Kronander, J., Denes, G., Mantiuk, R.K., Unger, J.: HDR image reconstruction from a single exposure using deep CNNs. ACM Trans. Graph. **36**(6) (2017). https://doi.org/10.1145/3130800.3130816

22. Jia, Y., et al.: Caffe: convolutional architecture for fast feature embedding. In: Proceedings of the 22nd ACM International Conference on Multimedia, MM 2014, pp. 675–678. Association for Computing Machinery, New York (2014). https://doi.org/10.1145/2647868.2654889

23. Kalal, Z., Mikolajczyk, K., Matas, J.: Tracking-learning-detection. IEEE Trans. Pattern Anal. Mach. Intell. **34**(7), 1409–1422 (2012)
24. Chen, Y., Schmid, C., Sminchisescu, C.: Self-supervised learning with geometric constraints in monocular video: connecting flow, depth, and camera. In: Proceedings of the 2019 IEEE/CVF International Conference on Computer Vision, pp. 7062–7071. IEEE, Piscataway (2019)
25. Wang, Z., Bovik, A.C., Sheikh, H.R., Simoncelli, E.P.: Image quality assessment: from error visibility to structural similarity. IEEE Trans. Image Process. **13**(4), 600–612 (2004). https://doi.org/10.1109/TIP.2003.819861
26. Ronneberger, O., Fischer, P., Brox, T.: U-Net: convolutional networks for biomedical image segmentation. In: Navab, N., Hornegger, J., Wells, W.M., Frangi, A.F. (eds.) MICCAI 2015. LNCS, vol. 9351, pp. 234–241. Springer, Cham (2015). https://doi.org/10.1007/978-3-319-24574-4_28

Image/Video Big Data Analysis and Understanding

Object-Aware Attention in Few-Shot Learning

Yeqing Shen[1], Lisha Mo[1], Huimin Ma[2(✉)], Tianyu Hu[2], and Yuhan Dong[1,3]

[1] Department of Electronic Engineering, Tsinghua University, Beijing, China
{shenyq18,mls18}@mails.tsinghua.edu.cn,
dongyuhan@sz.tsinghua.edu.cn
[2] Department of Computer and Communication Engineering, University
of Science and Technology Beijing, Beijing, China
{mhmpub,tianyu}@ustb.edu.cn
[3] Singhua Shenzhen International Graduate School, Shenzhen, China

Abstract. Embedding networks trained with a limited number of samples have a poor capability in localizing objects. Therefore, models of few-shot learning (FSL) are easily affected by object-irrelevant information in the background, which will lead to low accuracy. An Object-Aware Attention (OAA) mechanism is proposed to improve the generalization ability of models. In OAA module, object-relevant area is obtained by a fully convolutional network to guide the network in extracting object-relevant features. Besides, a general few-shot learning framework with OAA as a plug-and-play module is proposed, in which original images and object-aware images are fused to get the rectified prototypes. Under the general framework, the performance of most existing few-shot learning methods can be improved effectively. Comprehensive experiments show that the OAA can improve the accuracy of four mainstream baselines significantly. On benchmark *mini-ImageNet*, the method achieves a state-of-the-art performance on the 5-way-1-shot task and 5-way-5-shot task.

Keywords: Few-shot learning · Object-aware attention · Object-relevant features · Rectified prototypes

1 Introduction

Deep learning has made significant progress in image recognition in recent years. However, deep neural networks have weak scalability and poor performance with limited data, far inferior to human intelligence. It is a meaningful and interesting issue to learn through a limited number of samples, which is called Few-Shot Learning (FSL). Intuitively, it is expected that the embeddings are more relevant to the objects themselves and have a more considerable margin in inter-class variance than in intra-class. With sufficient training data, the performance of convolutional neural networks (CNN) is impressive on image recognition. Studies on class activation maps [1, 2] show that even though there is no location information of objects, the convolutional unit of each layer in CNN can still act as an object detector. The well-trained CNN with massive data can detect the bounding boxes of objects in the conventional image recognition task, while it is

© Springer Nature Singapore Pte Ltd. 2021
Y. Wang and W. Song (Eds.): IGTA 2021, CCIS 1480, pp. 95–108, 2021.
https://doi.org/10.1007/978-981-16-7189-0_8

difficult for an FSL model to obtain the location information of objects. In other words, due to the lack of data, the feature embeddings in few-shot learning models cannot be well concentrated on essential regions of objects, which will compromise the robustness of models.

For a one-shot learning task, there is only one support sample of each category available for training. Hence, the most discriminative dimensions in embeddings have a significant impact on the final classification boundary. If the background regions among different categories are similar, the object-irrelevant information will have a negative effect on the classification of query samples in the inference phase. As shown in Fig. 1, the query sample is a rabbit, with the same background as the dog in the support set. As Fig. 1(a) shows, the background of the query sample q is similar to that of the support sample s_2, while its object category is the same as the support sample s_3. In this case, the background is not related to the object, and the background can be regarded as object-irrelevant information. In the distance metric stage, object-irrelevant information introduced by backgrounds will cause interference to classification. As Fig. 1(b) shows, Object-Aware Attention (OAA) module replaces all samples with the same background, which helps the model focus on the inherent differences among these categories.

Transductive inferring mechanism [3–6] is proposed to obtain category prototypes with more category-relevant information. However, the prototypes produced by these transductive settings are dynamic, and the quality of prototypes depends on the query samples in the task. Therefore, an attention mechanism is established for few-shot learning to reduce the influence of irrelevant information and enhance the intrinsic object features. Some existing methods [7, 8] introduce the key-query attention mechanism into the few-shot learning framework to calculate the value matrix in order to weigh the different dimensions of feature vectors. Inspired by attention module design [9–11] in the fine-grained task, an OAA module is proposed in this paper to guide the embedding network in order to extract the features of the object-relevant regions and reduce background interference.

In order to obtain the localization of the object region, the saliency object detection algorithm is used in the OAA module to generate object-aware images. First, the original image is fed into a fully convolutional network (FCN). The encoders and down-sampling modules extract the high-level semantic features, and then the decoders and up-sampling modules generate saliency maps with different resolutions. Thus, the saliency maps in different resolutions are fused to obtain the final prediction, thereby obtaining object-aware images. After the object-aware image generation, two weight-shared networks are utilized to get original features and object-aware features, and then features from two branches are fused to get rectified prototypes. Since object-irrelevant features are weakened in object-aware images, the rectified prototypes will be closer to the essence of categories, reducing the interference information on the classification.

To compare the effect of background interference explicitly, a novel dataset *Animals5* is established to evaluate the robustness of existing methods, which consists of five kinds of animals with similar backgrounds. Main contributions of the paper are as following:

1. A novel Object-Aware Attention (OAA) module, which is efficient and plug-and-play, is proposed to reduce the influence of irrelevant information and enhance the object-relevant features.

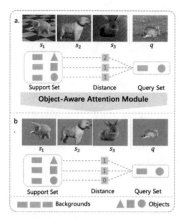

Fig. 1. The interference of object-irrelevant features. The combination of colors and shapes is used to simplify an image, representing background and object, respectively. At stage a, the background of q is similar to that of $s2$, and the object of q is similar with that of $s3$. This brings interference to classification. At stage b, backgrounds of all the samples are replaced with the same background with Object-Aware Attention module, which helps the model easily find the differences among categories.

2. A novel dataset *Animals5* with similar backgrounds is established for the first time, aiming to evaluate the models under the interference of background.
3. A two-branch general few-shot learning framework with the Object-Aware Attention module is constructed to efficiently compromise the robustness of irrelevant background interference and improve the performance of most existing methods.
4. We conduct comprehensive experiments on *mini-ImageNet* and *Animals5* and achieve a significant improvement over the state-of-the-art approaches. The model with OAA method has achieved higher state-of-the-art performance on 1-shot and 5-shot tasks on standard *mini-ImageNet* dataset.

2 Related Work

2.1 Few-Shot Learning

Existing few-shot learning methods can be categorized into two branches: meta-learning-based approaches and metric-learning based approaches.

On the one hand, meta-learning-based methods focus on updating parameters for a particular task. It aims to train a meta-learner on multiple few-shot tasks to expand the generalization ability on new tasks. In the meta-testing phase, the meta-learner model can be applied to new tasks with a few support samples. Typical methods MAML [12] aims to learn good initial parameters to guide models to adapt to new tasks in a few iterations. Reptile [13] is an improved version of MAML, and it retains first-order gradients based on MAML. To learn a task agnostic model, TAML [14] modifies the parameter updating formula of MAML and proposes the entropy-based method to learn an unbiased initial model in order to prevent over-performing in classification tasks. MAML and its variants

[12–17] have achieved impressive performance by optimizing model parameters, which guide the model adapt to new tasks quickly. In general, existing meta-learning methods train the learner with multiple tasks in the same mode and fine-tune it on new tasks. However, these methods are still prone to the interference of object-irrelevant features mentioned in Fig. 1.

On the other hand, metric-learning based methods focus on establishing a metric to calculate the similarity of samples. The similarity is generally defined as a distance function, such as Cosine or Euclidean distance in the embedding space. Most of the methods proposed in recent studies are based on metrics and can be subdivided into three types. The first type improves the performance by calculating and optimizing prototypes [3–6, 18–20]. Prototypical Network (PN) [18] regards the mean value of feature embeddings of each category as its prototype and assigns the query samples to the nearest prototype at the inference stage. The second type improves the performance with large-scale datasets, such as ImageNet [21], as auxiliary information [22, 23]. Typical method MTL [22] uses a large-scale dataset to train deep neural networks and adopts the meta-learning strategy to learn the parameters combined with fixed feature extracting layers. The third type improves the performance by utilizing Graph Neural Network (GNN) as a parameter update mechanism [24–27]. Classic GNN-based method EGNN [25] establishes an edge-labeling graph neural network that calculates the intraclass similarity and the inter-class dissimilarity by updating the edge labels.

In order to reduce the interference mentioned in Fig. 1, several methods rectify the prototype to contain less category-irrelevant information. For example, Category Transversal Module (CTM) proposed in [3] identifies category-relevant features based on both intra-class commonality and inter-class uniqueness in the embedding space. The embeddings of CTM reduce category-irrelevant embedding, thereby improving the prototype generalization and classification performance. MCT [4], CSPN [5] and LST [6] establish a transductive inferring mechanism similar to semi-supervised learning, which iteratively predicts the query samples and rectifies the prototypes in the query phase. Recently, transductive settings have been applied to many few-shot learning methods, which utilize the information of query samples to improve the performance of recognition. Inspired by the concept of transductive setting, the motivation of Object-Aware Attention proposed in this paper is to select category-relevant features from the prototype adaptively.

2.2 Saliency Object Detection

Saliency object detection aims to locate the most visually prominent object(s) in a given scene. Early studies integrate shallow features including color, edge, and texture, to locate salient regions. Some studies focus on predicting pixel-wise saliency maps, inspired by the fully convolutional networks. Wang et al. [28] produce saliency prior by low-level appearance cues and further apply it to rectify saliency prediction recurrently. Hou et al. [29] introduce short connections into fully convolutional networks to integrate features from different layers. R3net [30] and SRM [31] design strategies to iteratively refine the saliency maps step-by-step, with features from deep and shallow layers. Basnet [32] proposes a hybrid loss to use boundary information in prediction. These methods can achieve good performance on saliency object detection tasks, and can be used as possible solutions to obtain object-aware images in OAA.

U^2-Net [33] is a two-level nested U-structure. Units of the U-structure are fully-convolutional network proposed in [34], which can achieve the state-of-the-art performance on different segmentation applications with a limited number of training samples. In few-shot recognition, the object to be recognized usually occupies a prominent position in an image. Therefore, the saliency detection algorithm can locate the object area to reduce the interference of object-irrelevant features mentioned in Fig. 1.

3 Method

3.1 Preliminaries

In few-shot learning, the dataset is usually divided into two parts, training set $D_{train} = \{(x_i, y_i)|y_i \in Y_{train}\}$ and testing set $D_{test} = \{(x_i, y_i)|y_i \in Y_{test}\}$. Y_{train} and Y_{test} are sets of training and testing labels, and y_i is the label of sample x_i, and $Y_{train} \cap Y_{test} = \phi$. In meta-learning, the episodic strategy proposed in [1] is usually adopted to train the learner. First, D_{train} and D_{test} are randomly sampled to get episode sets $T_{train} = \{\tau_i | \tau_i \subset D_{train}\}$ and $T_{test} = \{\tau_i | \tau_i \subset D_{test}\}$, and every episode τ_i contains support set S and query set Q as Eq. (1) shows. For N-way-K-shot setting:

$$S = \left\{x_n^k\right\}, \ Q = \left\{x_n^m\right\}, \ n = 1, \ldots, \ N, k = 1, \ \ldots, K, \ m = K+1, \ldots, K+M \quad (1)$$

K and M are the number of samples of each category in the support set and query set, and N is the number of categories. That means each episode $\tau_i = (S_i, Q_i)$ has N categories and $K + M$ samples per category. For an episode τ, support set S_i is used to calculate the prototypes of the N categories, and to predict the category of each sample in Q_i. Then the loss is calculated in the training phase, or the accuracy is evaluated in the test phase. Since Y_{train} and Y_{test} are disjoint, this setting ensures that the model is only trained on K samples of each category in the testing phase.

3.2 Framework

A general few-shot learning framework with Object-Aware Attention module is proposed to reduce the interference of object-irrelevant features. As shown in Fig. 2, the framework comprises three stages: an Object-Aware Attention module to produce object-aware images, a two-branch structure with shared weights to extract feature embeddings, and a fusion module to combine the features of the two branches.

Stage I. For a recognition task, objects to be recognized are visually salient in images. Based on this assumption, an Object-Aware Attention module is designed with the idea of saliency object detection to obtain the object-aware region. A fully convolutional network [35] $FCN(\cdot)$ is trained with an auxiliary dataset to evaluate the significance of each pixel in the image. As shown in Eq. (2), saliency map is obtained by $FCN(\cdot)$, where each value represents the saliency of the pixel at that position, and then the object-aware image x^* is produced according to the saliency map. $\sigma_t(\cdot)$ is a characteristic function

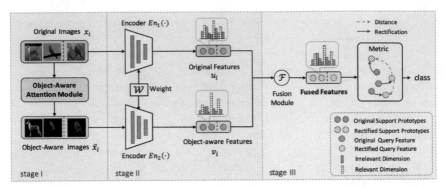

Fig. 2. General Few-shot Learning Framework with **Object-Aware Attention** Module. The framework consists of three stages. **Stage I:** Object-Aware images are obtained by the OAA module, in which object-irrelevant regions are replaced with the same pixels. **Stage II:** Images are input into two encoders respectively to obtain features, and the object-irrelevant features in object-aware features are effectively suppressed. **Stage III:** Original features and object-aware features are fused to calculate the rectified prototypes of categories (for support samples) and the rectified query features (for query samples), then the category of query features is predicted according to support prototypes in metric module.

that output 1 if the input larger than threshold t and output 0 if the input smaller than t. In experiments, t is set as 1.

$$x^* = \sigma_t(FCN(x)) \odot x \tag{2}$$

As shown in Fig. 2, the object-irrelevant pixels in the object-aware image are replaced with the same zero-value pixels, ensuring that the distinctive information of categories comes from the pixels of the object-relevant region. Specifically, the details on OAA module are introduced in Sect. 3.3.

Stage II. For example, there are two support samples x_1^{spt}, x_2^{spt} and one query sample x^{qry} in Fig. 2. At Stage I, object-aware images x_1^{*spt}, x_2^{*spt} and x^{*qry} are calculated. Then as Eq. (3) shows, x and x^* are fed into two encoders $En_1(\cdot)$ and $En_2(\cdot)$ respectively, which are weight-shared.

$$v_i = En_1(x_i), \ u_i = En_1\left(x_i^*\right) \tag{3}$$

It can be seen in Stage II of Fig. 2 that the original features v_i obtained by $En_1(\cdot)$ contain several dimensions, including object-relevant ones and object-irrelevant ones. In the upper branch, the values of irrelevant dimensions are close to the values of the relevant dimensions. As a result, the object-irrelevant information will be introduced when calculating category prototypes, which will lead to a poor generalization ability. In the bottom branch, it is obvious that the values of irrelevant dimensions are suppressed while the values of relevant dimensions are enhanced. Therefore, it helps reduce the intra-class variance and increase the inter-class distance.

Nevertheless, there are some situations where the Object-Aware Attention module cannot perceive good object-aware regions, such as chaotic backgrounds or camouflage colors. Hence, a two-branch structure with a feature fusion module is utilized as a trade-off to enhance object-relevant information.

Stage III. As Eq. (4) shows, original features and object-aware features are combined with the fusion module $F(\cdot)$, in which α is a hyperparameter and set to be 0.5. Compared with the original feature, the object-relevant dimensions after fusion are enhanced as expected. Hence, the category prototype calculated by the fused features contains essential object-relevant features while weakening the irrelevant information. The experimental results in the following section show that the rectified prototype has a better generalization ability.

$$F(v_i, u_i) = \alpha \cdot vi + (1 - \alpha) \cdot u_i \tag{4}$$

$$L_{cls} = L_{CE}(P(y_i|F(v_i, u_i), y_i)) \tag{5}$$

In the metric module, Euclidean distance is used to measure the similarity between prototype and query features. Besides, the cross-entropy loss function in Eq. (5) is used to iteratively update the model after predicting the category in the training phase.

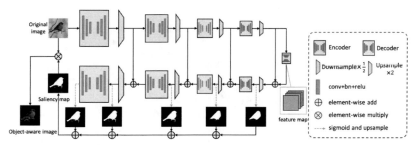

Fig. 3. The structure of object-aware attention module. The high-resolution original image is encoded to a low-resolution feature map. Then the feature map is reconstructed into high-resolution. Finally, saliency maps of different levels are combined together as the final saliency map to obtain the object-aware image.

3.3 Object-Aware Attention Module

The OAA module utilizes the idea of saliency object detection to locate objects in the image in order to obtain object-aware images. Inspired by [33], network in OAA is a fully convolutional network consisting of symmetrical encoders and decoders. The network is able to capture more contextual information from different scales with the mixture of receptive fields, and the network can capture richer local and global information with the nested symmetry with residual U-block. Actually, any saliency object detection method with a good performance can be applied instead. As shown in Fig. 3, the input is fed into the encoders and the down-sample modules, and the high-resolution original image is encoded to a low-resolution feature map. Then the feature map is reconstructed into high-resolution by the decoders and up-sample modules.

The output of the encoder and that of the decoder with the same resolution are combined to obtain the intermediate saliency map. Several saliency maps are combined with a saliency map fusion module. In order to obtain the saliency detection result with better resolution, the encoder and decoder are composed of symmetrical convolutional layers. Besides, the high-resolution features are combined with the up-sampled output to get a more accurate output. Supervision information mentioned in [33] is used to train the network on the dataset DUTS-TR [36], which can help to get the localization information of the objects in the image. Finally, the original image is element-wised multiplied with the saliency map to generate the object-aware image.

4 Experiments

4.1 Datasets

- *mini-ImageNet* was initially proposed in [37] and consists of 100 categories, which are randomly chosen from ILSVRC2012 [21]. The 100 categories are divided into three parts: 64 for training, 16 for validation and 20 for testing. Each category contains 600 RGB-colored images with the size of 84 × 84.

Fig. 4. *Animals5* dataset examples. The dataset includes 1204 images of 5 animal categories, with only one object in each image. The background of images in *Animals5* is grassland or white wall, which can be used to explore the impact of similar backgrounds on algorithm performance.

- *Animals5* is proposed here to evaluate the robustness of methods under similar background interference. The dataset is constructed by online images according to category keywords. The dataset includes 1204 images of 5 animal categories, with only one object in each image. Specifically, there are 257 images of cats, 276 images of dogs, 236 images of rabbits, 211 images of cows and 224 images of sheep. The background of images in *Animals5* dataset is grassland or white wall, which can be used to explore the impact of similar backgrounds on algorithm performance. Examples of *Animals5* dataset are demonstrated in Fig. 4.

4.2   Baseline Models and Experiment Details

- *MAML* [12] learns to search for the optimal initialization to fast adapt to a new few-shot task. A lightweight network which contains four groups of Conv-Bn-Relu units (a unit contains a convolution layer, a batch normalization layer and a Relu layer) is used in MAML. The initial learning rate of network is set as 0.001, and the learning rate of the specific episode is set as 0.01. Model trains within 6 epochs and each epoch includes10kepisodes.

- *PN* [18] defines the mean embedding of every class as category prototypes, and the query samples are assigned to their nearest prototype in the test phase. The same network is applied as used in MAML [12], and the initial learning rate is set to 0.001, which decays by 0.5 times every 20epochs. Model trains within 200 epochs and each epoch is consisted of 100 episodes.

- *MCT* [4] meta-learns the confidence for each query sample to assign optimal weights to unlabeled queries such that they improve the transductive inference performance of the model on unseen tasks. In order to simplify the model and better explore the impact of OAA on the performance of the model, the dense classification loss of MCT [4] is removed and the method is denoted as w-MCT. A classicResNet-12 network as mentioned in [4] is applied, and the model is trained with 50k episodes in one epoch. The learning rate is set to 0.1 at first, and it decays to 0.006and 0.0012 at 25k-th episode and 35k-th respectively (Fig. 5).

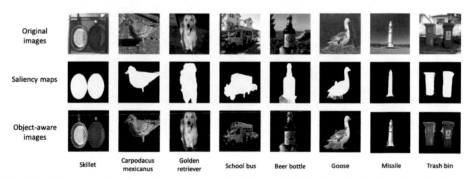

Fig. 5. Object-aware images of several categories in ImageNet. The first line is the original images, the second line is the saliency maps, and the third line is the object-aware images.

MAML [12] and PN [18] are two representative methods based on meta-learning and metric-learning, respectively. Besides, MCT [4] is also a representative method with transductive inference and achieves state-of-the-art results on several benchmark datasets. Therefore, these three methods are chose as baselines.

4.3   Experiment Result

Following previous work, all classification accuracy results are averaged over 600 test tasks. The results in Table 1 compare the performance of some existing methods and 4 baselines with the OAA module on *mini-ImageNet*. Two settings are adopted to evaluate OAA module, denoted as $(\cdot)+$ and $(\cdot)++$ respectively.

The first setting is to use OAA module only in support set, which helps to evaluate the generalization ability of the prototypes extracted by the model after adding OAA. Methods denoted as $(\cdot)+$ in Table 1 are under this setting. For the two tasks, i.e., 5-way-1-shot and 5-way-5-shot, OAA module under this setting improves the performance by 0.27/0.54, 0.11/0.22, 2.29/0.85 and 0.97/0.74 points on four baselines (MAML, PN, w-MCT and MCT), respectively. The results show that after adding OAA to support samples, the final accuracy rate is significantly improved, which means that the rectified prototypes are more object-relevant, and the generalization ability is stronger. This shows

Table 1. Classification results on *mini-ImageNet* under 5-way-1-shot and 5-way-5-shot setting. Methods denoted as $(\cdot)^*$ are the results that we reimplement. Methods denoted as $(\cdot)+$ means OAA is only implemented on support set. Methods denoted as $(\cdot)++$ means OAA is implemented on both support set and query set. w-MCT denotes MCT without dense classification loss.

Method	5-way-1-shot Acc	5-way-5-shot Acc
MatchingNet [37]	46.60 ± 0.84	55.31 ± 0.73
Reptile [13]	49.97 ± 0.32	65.99 ± 0.58
BAN [7]	53.74 ± 0.89	71.90 ± 0.76
TPN [19]	52.78 ± 0.27	66.59 ± 0.28
HPN [38]	55.17 ± 0.61	71.26 ± 0.69
FGNN [27]	64.15 ± 0.28	80.08 ± 0.35
CAN [40]	67.19 ± 0.55	80.64 ± 0.35
MAML [12]	48.70 ± 1.84	63.11 ± 0.92
MAML+(ours)	48.97 ± 1.43	63.65 ± 0.98
MAML++(ours)	**49.40 ± 1.22**	**63.92 ± 1.01**
PN* [18]	49.21 ± 0.38	68.20 ± 0.66
PN+(ours)	49.32 ± 0.40	68.42 ± 0.49
PN++(ours)	**50.84 ± 0.37**	**71.34 ± 0.57**
w-MCT* [4]	63.62 ± 0.43	74.94 ± 0.55
w-MCT+(ours)	65.91 ± 0.40	75.79 ± 0.52
w-MCT++(ours)	**68.59 ± 0.26**	**77.00 ± 0.59**
MCT [4]	76.16 ± 0.89	85.22 ± 0.42
MCT+(ours)	77.13 ± 0.41	85.96 ± 0.44
MCT++(ours)	**77.16 ± 0.37**	**86.67 ± 0.49**

that excluding irrelevant information in the background is helpful for the final object recognition.

The other approach is to use OAA module in both support set and query set, and these methods are denoted as $(\cdot)$++. For the two tasks, i.e., 5-way-1-shot and 5-way-5-shot, OAA module under this setting improves the performance by 0.70/0.81, 1.63/3.14, 4.97/2.06 and 1.00/1.45 points on four baselines (MAML, PN, w-MCT and MCT), respectively.

Table 2. Classification results on *Animals5* under 5-way-1-shot and 5-way-5-shot setting. Methods denoted as $(\cdot)^*$ are the results that we reimplement. Methods denoted as $(\cdot)$+ means OAA is only implemented on support set. Methods denoted as $(\cdot)$++ means OAA is implemented on both support set and query set. w-MCT denotes MCT without dense classification loss.

Method	5-way-1-shot Acc	5-way-5-shot Acc
MAML[12]	21.52 ± 1.36	25.44 ± 0.98
MAML+(ours)	$\mathbf{28.40 \pm 1.10}$	$\mathbf{35.08 \pm 0.84}$
MAML++(ours)	26.68 ± 1.02	33.70 ± 0.94
PN* [18]	36.11 ± 0.18	44.05 ± 0.28
PN+(ours)	$\mathbf{37.03 \pm 0.20}$	$\mathbf{45.10 \pm 0.31}$
PN++(ours)	36.67 ± 0.24	44.70 ± 0.35
w-MCT* [4]	33.86 ± 0.21	44.15 ± 0.39
w-MCT+(ours)	$\mathbf{34.22 \pm 0.27}$	$\mathbf{45.11 \pm 0.46}$
w-MCT++(ours)	33.66 ± 0.19	44.50 ± 0.38
MCT [4]	33.72 ± 0.25	43.36 ± 0.46
MCT+(ours)	$\mathbf{34.60 \pm 0.20}$	$\mathbf{45.01 \pm 0.44}$
MCT++(ours)	34.36 ± 0.18	44.34 ± 0.48

It can be seen that under this setting, OAA brings a higher improvement to the baselines, which means applying OAA on both the support set and the query set can effectively reduce the intra-category variance and help the network extract common features within the category. It is worth emphasizing that OAA module is able to improve the performance of w-MCT with relatively accuracy gain of nearly 5 points. For the two tasks, i.e., 5-way-1-shot and 5-way-5-shot, MCT++ succeed in achieving a higher state-of-the-art performance at 77.16% and 86.67% on *mini-ImageNet* dataset.

The experiment results on the *Animals5* are directly obtained with the model trained on the *mini-ImageNet*, which is in line with the setting of meta-learning [12]. It notes that *Animals5* accuracy in Table 2 is obviously lower than *mini-ImageNet* accuracy, which shows that the existing baselines are vulnerable to interference from similar backgrounds. After applying OAA, the model learns to pay attention to object-aware features, thus the accuracy of $(\cdot)$+ and $(\cdot)$++ has been significantly improved. This illustrates the importance of finding object-relevant features in few-shot learning research. The rectified prototypes contain less category-irrelevant information, which can be applied to reduce the interference in the background.

5 Conclusion

In this paper, an efficient and plug-and-play Object-Aware Attention module is proposed to reduce the influence of object-irrelevant information and enhance the object-aware features. Besides, the *Animals5* dataset is established to evaluate models under similar background interference. Experiments results show that OAA has the ability to reduce the interference of irrelevant information and extract object-relevant intrinsic information. With OAA, few-shot learning methods are able to focus on the feature of the object. The general framework with OAA has improved the performance of most existing few-shot learning approaches efficiently and promoted the robustness and generalization ability of these models.

Acknowledgment. This work was supported by the National Natural Science Foundation of China (No. U20B2062), the Beijing Municipal Science \& Technology Project (No. Z191100007419001), the Beijing National Research Center for Information Science and Technology, and the key Laboratory of Opto-Electronic Information Processing, CAS (No. JGA202004027).

References

1. Zhou, B., Khosla, A., Lapedriza, A., Oliva, A., Torralba, A.: Learning deep features for discriminative localization. In: Proceedings of the IEEE Conference on Computer Vision and Pattern Recognition, pp. 2921–2929. IEEE Press, New York (2016)
2. Zhou, Y., Zhu, Y., Ye, Q., Qiu, Q., Jiao, J.: Weakly supervised instance segmentation using class peak response. In: Proceedings of the IEEE Conference on Computer Vision and Pattern Recognition, pp. 3791–3800. IEEE Press, New York (2018)
3. Li, H., Eigen, D., Dodge, S., Zeiler, M., Wang, X.: Finding task-relevant features for few-shot learning by category traversal. In: Proceedings of the IEEE Conference on Computer Vision and Pattern Recognition, pp. 1–10. IEEE Press, New York (2019)
4. Kye, S.M., Lee, H.B., Kim, H., Hwang, S.J.: Meta-learned confidence for few-shot learning. arXiv e-prints, arXiv-2002 (2020)
5. Liu, J., Song, L., Qin, Y.: Prototype rectification for few-shot learning. arXiv preprint arXiv: 1911.10713 (2019)
6. Li, X., et al.: Learning to self-train for semi-supervised few-shot classification. In: NeurIPS, vol. 32, pp. 10276–10286 (2019)
7. Ke, L., Pan, M., Wen, W., Li, D.: Compare learning: bi-attention network for few-shot learning. In: ICASSP 2020-2020 IEEE International Conference on Acoustics, Speech and Signal Processing, pp. 2233–2237. IEEE Press, New York (2020)
8. Li, R., Liu, H., Zhu, Y., Bai, Z.: Arnet: attention-based refinement network for few-shot semantic segmentation. In: ICASSP 2020-2020 IEEE International Conference on Acoustics, Speech and Signal Processing, pp. 2238–2242. IEEE Press, New York (2020)
9. Fu, J., Zheng, H., Mei, T.: Look closer to see better: recurrent attention convolutional neural network for fine-grained image recognition. In: Proceedings of the IEEE Conference on Computer Vision and Pattern Recognition, pp. 4476–4484. IEEE Press, New York (2017)

10. Zheng, H., Fu, J., Mei, T., Luo, J.: Learning multi-attention convolutional neural network for fine-grained image recognition. In: Proceedings of the IEEE International Conference on Computer Vision, pp. 5209–5217. IEEE Press, New York (2017)

11. Wang, Y., Mo, L., Luo, X., Ma, H.: Essential element-region driven model in image recognition. Neurocomputing **349**, 116–122 (2019)

12. Finn, C., Abbeel, P., Levine, S.: Model-agnostic meta-learning for fast adaptation of deep networks. In: International Conference on Machine Learning, pp. 1126–1135. PMLR (2017)

13. Nichol, A., Achiam, J., Schulman, J.: On first-order meta-learning algorithms. arXiv preprint arXiv:1803.02999 (2018)

14. Jamal, M.A., Qi, G.J.: Task agnostic meta-learning for few-shot learning. In: Proceedings of the IEEE/CVF Conference on Computer Vision and Pattern Recognition, pp. 11719–11727. IEEE Press, New York (2019)

15. Santoro, A., et al.: A simple neural network module for relational reasoning. arXiv preprint arXiv:1706.01427 (2017)

16. Lee, K., Maji, S., Ravichandran, A., Soatto, S.: Meta-learning with differentiable convex optimization. In: Proceedings of the IEEE/CVF Conference on Computer Vision and Pattern Recognition, pp. 10657–10665. IEEE Press, New York (2019)

17. Rusu, A.A., et al.: Meta-learning with latent embedding optimization. arXiv preprint arXiv:1807.05960 (2018)

18. Snell, J., Swersky, K., Zemel, R.S.: Prototypical networks for few-shot learning. In: Proceedings of the 31st International Conference on Neural Information Processing Systems, pp. 4080–4090. Red Hook, New York (2017)

19. Liu, Y., et al.: Learning to propagate labels: transductive propagation network for few-shot learning. arXiv preprint arXiv:1805.10002 (2019)

20. Zhang, X., Qiang, Y., Sung, F., Yang, Y., Hospedales, T.: RelationNet2: deep comparison network for few-shot learning. In: 2020 International Joint Conference on Neural Networks, pp. 1–8. IEEE Press, New York (2020)

21. Krizhevsky, A., Sutskever, I., Hinton, G.E.: Imagenet classification with deep convolutional neural networks. In: Proceedings of the 25th International Conference on Neural Information Processing Systems, pp. 1097–1105. IEEE Press, New York (2012)

22. Sun, Q., Liu, Y., Chua, T.S., Schiele, B.: Meta-transfer learning for few-shot learning. In: Proceedings of the IEEE/CVF Conference on Computer Vision and Pattern Recognition, pp. 403–412. IEEE Press, New York (2019)

23. Liu, L., Zhou, T., Long, G., Jiang, J., Yao, L., Zhang, C.: Prototype propagation networks (PPN) for weakly-supervised few-shot learning on category graph. arXiv preprint arXiv:1905.04042 (2019)

24. Garcia, V., Bruna, J.: Few-shot learning with graph neural networks. arXiv preprint arXiv:1711.04043 (2017)

25. Kim, J., Kim, T., Kim, S., Yoo, C.D.: Edge-labeling graph neural network for few-shot learning. In: Proceedings of the IEEE/CVF Conference on Computer Vision and Pattern Recognition, pp. 11–20. IEEE Press, New York (2019)

26. Gidaris, S., Komodakis, N.: Generating classification weights with GNN denoising autoencoders for few-shot learning. In: Proceedings of the IEEE/CVF Conference on Computer Vision and Pattern Recognition, pp. 21–30. IEEE Press, New York (2019)

27. Wei, T., Hou, J., Feng, R.: Fuzzy graph neural network for few-shot learning. In: 2020 International Joint Conference on Neural Networks, pp. 1–8. IEEE Press, New York (2020)

28. Wang, L., Wang, L., Lu, H., Zhang, P., Ruan, X.: Saliency detection with recurrent fully convolutional networks. In: Leibe, B., Matas, J., Sebe, N., Welling, M. (eds.) ECCV 2016. LNCS, vol. 9908, pp. 825–841. Springer, Cham (2016). https://doi.org/10.1007/978-3-319-46493-0_50

29. Hou, Q., Cheng, M.M., Hu, X., Borji, A., Tu, Z., Torr, P.H.: Deeply supervised salient object detection with short connections. In: Proceedings of the IEEE Conference on Computer Vision and Pattern Recognition, pp. 3203–3212. IEEE Press, New York (2017)

30. Deng, Z., et al.: R3net: recurrent residual refinement network for saliency detection. In: Proceedings of the 27th International Joint Conference on Artificial Intelligence, pp. 684–690. AAAI Press, Palo Alto (2018)

31. Wang, T., Borji, A., Zhang, L., Zhang, P., Lu, H.: A Stagewise refinement model for detecting salient objects in images. In: Proceedings of the IEEE International Conference on Computer Vision, pp. 4019–4028. IEEE Press, New York (2017)

32. Qin, X., Zhang, Z., Huang, C., Gao, C., Dehghan, M., Jagersand, M.: BASNet: boundary-aware salient object detection. In: Proceedings of the IEEE/CVF Conference on Computer Vision and Pattern Recognition, pp. 7479–7489. IEEE Press, New York (2019)

33. Qin, X., Zhang, Z., Huang, C., Dehghan, M., Zaiane, O.R., Jagersand, M.: U2-net: going deeper with nested U-structure for salient object detection. Pattern Recognit. **106**, 107404 (2020)

34. Ronneberger, O., Fischer, P., Brox, T.: U-net: convolutional networks for biomedical image segmentation. In: Navab, N., Hornegger, J., Wells, W., Frangi, A. (eds.) MICCAI 2015. LNCS, vol. 9351, pp. 234–241. Springer, Cham (2015). https://doi.org/10.1007/978-3-319-24574-4_28

35. Shelhamer, E., Long, J., Darrell, T.: Fully convolutional networks for semantic segmentation. In: IEEE Transactions on Pattern Analysis and Machine Intelligence, pp. 640–651. IEEE, New York (2017)

36. Wang, L., et al.: Learning to detect salient objects with image-level supervision. In: Proceedings of the IEEE Conference on Computer Vision and Pattern Recognition, pp. 136--145. IEEE Press, New York (2017)

37. Vinyals, O., Blundell, C., Lillicrap, T., Kavukcuoglu, K., Wierstra, D.: Matching networks for one shot learning. In: Proceedings of the 30th International Conference on Neural Information Processing Systems, pp. 3637–3645. Red Hook, New York (2016)

38. Tan, S., Yang, R.: Hybrid pooling networks for few-shot learning. In: 2020 International Joint Conference on Neural Networks, pp. 1–8. IEEE Press, New York (2020)

39. Zhang, F., Wang, Q., Li, X.: Deep meta-relation network for visual few-shot learning. In: ICASSP 2020-2020 IEEE International Conference on Acoustics, Speech and Signal Processing, pp. 1509–1513. IEEE Press, New York (2020)

40. Hou, R., Chang, H., Ma, B., Shan, S., Chen, X.: Cross attention network for few-shot classification. arXiv preprint arXiv:1910.07677 (2019)

Simultaneously Predicting Video Object Segmentation and Optical Flow Without Motion Annotations

Jingchun Cheng[1], Shengjin Wang[2], and Chunxi Zhang[1(✉)]

[1] Beihang University, Beijing, China
zhangchunxi@buaa.edu.cn
[2] Tsinghua University, Beijing, China
wgsgj@tsinghua.edu.cn

Abstract. Optical flow information is one of the most commonly used temporal cues in video object segmentation algorithms. However, as it is difficult to label real-world video data with motion annotations, video segmentation methods are often forced to use external optical flow datasets and additional flow prediction models. In this paper, we propose an optical flow synthesizing approach which can generate artificial object flow from video segmentation masks, relieving the constraint of manual motion annotations for joint learning of video segmentation and optical flow prediction tasks. Extensive experiments and analysis are carried out on the DAVIS video segmentation datasets and the self-constructed synthetic flow database, demonstrating that the proposed synthetic flow has a better training effect compared with external flow datasets, and that this target-specific flow synthesizing training scheme can help video segmentation networks to better distinguish the motion patterns of certain targets in multiple-instance video segmentation scenes.

Keywords: Object flow · Target-specific flow synthesizing training · Joint learning and single/multiple instance video object segmentation

1 Introduction

Proposed in recent years, video object segmentation is a challenging video processing task which requires methods to track and segment a target instance throughout video frames pixel-wisely with very little prior knowledge. It has attracted much research interest due to its wide range of potential applications, e.g. autonomous driving [4,10,26], video editing [8,30] and human-machine interaction [1,5].

Video object segmentation tasks can be categorized into 'unsupervised' and 'semi-supervised' depending on whether an initial object mask is given in the test video. 'Semi-supervised' video segmentation task can further be classified into 'single-instance' and 'multiple-instance' settings by the number of target objects. Note that for the 'unsupervised' video object segmentation task, the

© Springer Nature Singapore Pte Ltd. 2021
Y. Wang and W. Song (Eds.): IGTA 2021, CCIS 1480, pp. 109–124, 2021.
https://doi.org/10.1007/978-981-16-7189-0_9

| Frame t | Frame t+1 | Optical Flow | Object Flow |

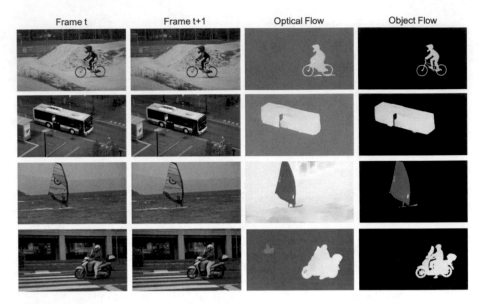

Fig. 1. Comparison of real-world and synthetic flow on the DAVIS 2016 dataset. This figure shows the real-world optical flow data (3rd column) and the proposed synthetic object flow (4th column). Note that the DAVIS 2016 dataset does not have motion annotations, therefore its real-world optical flow is approximately represented by the prediction results of [12]. We can see that the proposed synthetic flow is similar to real-world flow data.

test video should only contain one primary instance; otherwise, methods can not differentiate the target instance from other objects.

Numerous algorithms have been developed to tackle these tasks. As a test video provides no more than one glimpse of the initial target appearance with unknown object category, methods extensively exploit the available information in space and temporal dimensions. There are the appearance-based modules [6,23,29,34,37] who segment each frame independently, recognizing target instance only by its appearance feature; and the propagation-based modules [15,36] who process videos frame by frame, utilizing the temporal flow connections to track instances. For example, the appearance-based module [34] mainly relies on the extraordinary representation ability of CNN: it fine-tunes specific networks on the initial frames, and then segments the test video frame by frame, independent from other information; while the propagation-based module [15] guides previous segmentation masks by optical flow to segment the next frame. The mainstream of state-of-the-art video object segmentation methods use both the appearance and temporal information, e.g. [13] takes in previous frames as well as the optical flow amplitude maps and trains a network to link and group segmentation proposals along video frames. Similar to [13], most such methods incorporate the video temporal information in the form of optical flow, which contains rich and dense motion context between adjacent frames, i.e. the pixel-wise coordinate correspondence. Figure 1 provides some examples of the optical flow motion maps in real-world videos, illustrating that optical flow is very informative for video segmentation (pixels within one target tend to have the similar optical flow).

However, flow estimation itself is a challenging problem. Inaccurate flow information may harm segmentation performance instead of improving it. Moreover, video segmentation datasets are composed of real-world videos without ground truth optical flow labels, making it hard for methods to learn the exact motion pattern for each video target. Therefore, most existing methods are forced to use additional optical flow prediction models trained upon external virtual flow datasets (like the SINTEL dataset [2], the SceneFlow datasets [22]), resulting in a complex and redundant segmentation framework. Only a few joint learning based methods have simpler frameworks without upfront flow estimation modules while maintaining high performance. For example, the *SegFlow* [7] network is a unified, end-to-end trainable convolutional neural network for video object segmentation which can jointly predict segmentation mask and optical flow in videos. Figure 2 shows the network structure of *SegFlow*, where we can see that it has a straightforward prediction framework which only needs video frames as input. However, even such clean networks face the problem of lacking motion annotations. [7] smartly deals with this problem by iteratively switching through one dataset with segmentation annotations and another one with flow annotations during training. The iterative training scheme does enable the network to predict optical flow (Fig. 2), but its flow estimation confuses segmentation prediction when the video has a dynamic background as warned itself in [7]. Another drawback of training with an irrelevant flow database is that without a specific target motion pattern, *SegFlow* can only extract the global motion feature even in the 'semi-supervised' setting and cannot be applied to multiple-instance video segmentation tasks.

As described above, there is a lack of datasets with both segmentation and optical flow annotations. This is because that optical flow can be computed in a simulation scene where objects and background are all in control, but is very complicated to measure in the real-world situation. As a result, video object segmentation algorithms which require motion information often need to include extra data and estimation results from some vastly different virtual optical flow datasets. To tackle this problem, we propose an object flow synthesizing method which can produce virtual object motion in video segmentation datasets. The proposed synthesizing step can generate target movement and its corresponding flow ground truth from the segmentation images and annotations (examples of the generated flow are shown in Fig. 1, column 4). We prove that our synthetic flow data can resemble real-world object movements and provide infinite training data once given the segmentation dataset. We use *SegFlow* as an example model, showing that it can be optimized favorably with only one video segmentation dataset via the proposed training scheme. Besides, we also provide novel evaluation criteria to quantitatively measure the flow quality in videos with segmentation masks as reference.

The closest to our work is [17], which focuses on data synthesizing to help train networks on video object segmentation datasets. However, the data synthe-

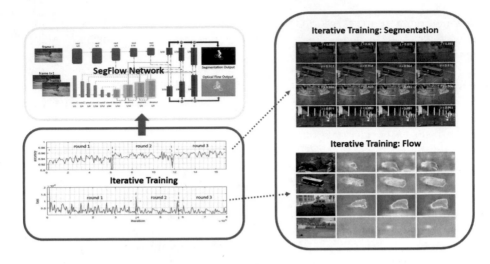

Fig. 2. Framework of the *SegFlow* joint prediction model. The network has a dual-branch structure (top-left corner), and is optimized iteratively between two datasets for three rounds (example results from each round are shown on the right).

sizing process in [17] is very complicated: they generate a large amount of training data per video by cutting out object area, inpainting background area, transforming both areas and recomposing them to obtain the final result. Although [17] proves that their network can achieve state-of-the-art performance with only the synthetic data, with the complexity in each of their data synthesizing steps, it needs massive time (half an hour or so) for data preparation per test video before training networks. In addition, the ground truth of their synthetic optical flow stays unused in the training and testing process. In this paper, we propose to use a simple and fast way to generate object flow data and its corresponding ground truth. We also show that with this synthetic data, we can train segmentation as well as optical flow predictions using a single video segmentation dataset, which is one step further than *SegFlow* [7] that requires an additional optical flow dataset.

To demonstrate the effectiveness of the proposed network, we carry out extensive experiments on both the single-instance and multiple-instance video object segmentation tasks [24,25], as well as the self-constructed virtual flow dataset. The contributions of this work are as follows:

- we propose an object flow synthesizing algorithm, which can generate object movement and corresponding ground truth from the segmentation annotation in a fast and simple way, producing object motion and labels for video segmentation target(s);
- we train the joint-learning network, *SegFlow*, with the proposed synthetic flow data, validating that the proposed flow synthesizing training scheme is able to train segmentation and flow joint prediction models without manual motion annotations;

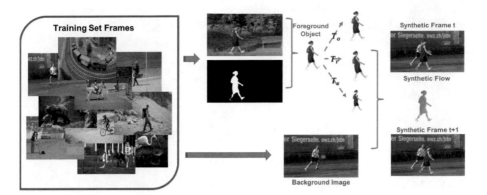

Fig. 3. Object flow synthesizing process. We randomly select two images from the training set, from which one is used as a static background image, and the object in the other one is cut out and transformed via transformation, rotation and scaling before being embedded to the background image. The object flow ground truth is obtained via transformation matrix defined in Sect. 2.2.

– the single dataset trained *SegFlow* performs better than the one trained from multiple datasets, demonstrating the effectiveness of the proposed training scheme.

2 Methodology

2.1 Base Model

We use the video object segmentation and optical flow joint prediction model, *SegFlow* as the base model to validate the proposed data synthesizing and training scheme. The *SegFlow* is a dual-branch, end-to-end convolutional neural network which is inspired by the effectiveness of fully-convolutional networks in image segmentation [20] and the deep structures in image classification [11,28]. It consists of a segmentation branch and an optical flow estimation branch, whose features in different convolution levels densely interact and fuse with each other. The original training scheme of *SegFlow* is to freeze one branch and train the other branch on a corresponding dataset (i.e. the DAVIS dataset for segmentation branch, and the SINTEL dataset for optical flow branch). After about three rounds of iterative training, the network converges to a joint optimization point where both segmentation and flow predictions are favorable (Fig. 2).

Instead of using another flow dataset and iteratively switching between two different kinds of data and labels, in this paper, we optimize the *SegFlow* with only the video object segmentation dataset, simultaneously training the network with both the segmentation and the proposed synthetic flow.

2.2 Object Flow Synthesizing

As mentioned above, there is a lack of large datasets with both annotations of segmentation mask and optical flow, making it difficult for algorithms to exploit

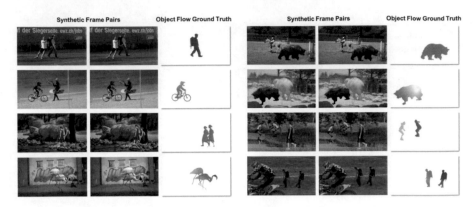

Fig. 4. Examples of the synthetic flow data generated from the DAVIS segmentation datasets with single target and multiple targets.

the combination of these two features. Besides, current optical flow methods are mostly trained on external virtual data [2,22], where objects and surroundings are very different from real-world videos in video object segmentation datasets [24,25,38]. Based on this observation, we propose an object flow synthesizing algorithm which can imitate the movement of any video object in the segmentation dataset. Our synthesizing method is simple, fast and can effectively provide target motion patterns for video object segmentation datasets. We show that with the synthetic flow data, the base net *SegFlow* can be trained with only video segmentation datasets, which is more convenient than [7] that requires to train its flow branch on the SINTEL flow dataset. We also demonstrate in Sect. 4 that *SegFlow* can obtain even better segmentation results with this synthetic flow training scheme.

Object Flow Computation. During synthesizing, we focus on object movement in videos and ignore other environmental factors. To generate more realistic object flow on videos with segmentation labels, we first observe the optical flow estimation in object area by state-of-the-art algorithms [12]. Figure 1 shows some examples of the flow estimation, from which we find that most object movements are quite simple, and resemble affine transformation, e.g. translation, rotation, scaling. Therefore, we can manually move, rotate and resize the video objects and calculate their corresponding flow with formula (1):

$$[u, v, 0] = [h, w, 1] * (T - I), \tag{1}$$

where $[u, v]$ denotes the optical flow at position $[h, w]$ in the image, and T denotes the transformation matrix in Eq. (2). Note that T integrates the transformation matrix T_o, scaling matrix T_s and rotation matrix T_r:

$$T = T_o * T_s * T_r, \tag{2}$$

$$T_o = \begin{bmatrix} 1 & 0 & h_{off} \\ 0 & 1 & w_{off} \\ 0 & 0 & 1 \end{bmatrix}, \tag{3}$$

$$T_s = \begin{bmatrix} 1+sc & 0 & 0 \\ 0 & 1+sc & 0 \\ 0 & 0 & 1 \end{bmatrix}, \tag{4}$$

$$T_r = \begin{bmatrix} cos\theta & -sin\theta & 0 \\ sin\theta & cos\theta & 0 \\ 0 & 0 & 1 \end{bmatrix}. \tag{5}$$

The parameters w_{off}, h_{off} denote the horizontal and vertical object translation respectively, sc denotes the scaling factor, and θ controls the rotation angle.

Synthesis Process. We generate two kinds of object flow, i.e. with single moving object and multiple moving objects respectively.

- Single Object. We randomly select one frame (I_{bg}) from training videos as background, and another frame (I_{fg}) with its object (area of foreground annotation larger than 100) as foreground. The foreground object is cut from I_{fg} using the segmentation annotation, and added to a random position in I_{bg} as the initial image. We keep a static background, and assign a minor movement to the foreground object using the transformation matrix described above. Then we obtain a synthetic frame and its optical flow ground truth by adding the transformed object to the same position in I_{fg}. Figure 3 shows the pipeline of our synthesizing process.
- Multiple Objects. Similarly, we randomly select one frame (I_{bg}) from training videos as background, and another frame (I_{fg}) with objects (from videos with instance-level segmentation annotations) as foreground. We move one object at a time by adding the extracted object and its transformation format to I_{bg}. We ensure that objects do not occlude each other to a large extent by calculating the overlap rate before adding to I_{bg} (we stop adding the current object if it overlaps too much with previous ones). Figure 4 presents some examples of our synthetic flow data with single or multiple objects.

3   Network Implementation and Training

Training of the *SegFlow* network have two major steps: offline training and online training, where the offline training step uses all training videos to train a generic segmentation model that can segment all foreground video targets; and the online training step uses one annotated test video frame to fine-tune a specific model that locates onto a particular video target.

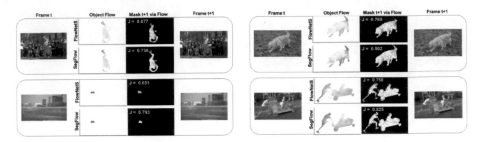

Fig. 5. Visualization of the flow prediction results and the masks warped via flow results on the DAVIS dataset.

3.1 Offline Training

For the offline training, the original *SegFlow* needs to iteratively train between two separate datasets with segmentation and optical flow annotations, respectively. With the synthetic object flow, we can train the *SegFlow* using only the video object segmentation dataset: we pretrain the flow branch upon our self-constructed flow dataset, and then feed pairs of synthetic frames with corresponding motion labels to train the joint network. During training, we use the cross-entropy loss for the segmentation branch, and the end-point-error loss for the optical flow branch; both branches are optimized simultaneously. We set the learning rate to be 1e−6, and train for about 100'000 iterations until the two losses both reach convergence.

3.2 Online Training

For the online training, without the synthetic flow, the original *SegFlow* model updates only the segmentation branch from the initial frame; its flow branch still provides motion representations to segmentation, but its parameters do not update. In contrast, our object flow synthesizing algorithm enables *SegFlow* to update optical flow branch parameters as well in the online updating process. We randomly generate synthetic flow using the initial frame as I_{fg} and I_{bg} (see Sect. 2.2 for details), and update both branches with a fixed learning rate of 1e−8. As a result, the network learns from the information of both flow and segmentation and has a better cognition of target instances.

One important difference is that the *SegFlow* without the synthetic flow data cannot tell the difference among several object motions in multiple-instance segmentation scenes, and can only be applied to single-instance video object segmentation task. When training with the synthetic flow, the network is provided with the specific object motion pattern and can therefore train its flow branch to focus on one instance, enabling the network to distinguish different targets in the multiple-instance video object segmentation task.

Table 1. Ablation study on the DAVIS 2016 validation set. We show performance of *SegFlow* with different training settings, i.e. removing the online-training, flow branch, synthetic flow in the offline (-SynFlow offline) and online (-SynFlow online) training process.

Method	Online training	Motion annotation	Flow branch	J mean	F mean	T mean
Ours	✓		✓	0.770	0.766	0.223
-SynFlow online	✓		✓	0.749	0.746	0.240
-flow branch	✓			0.720	0.720	0.251
Ours offline			✓	0.667	0.667	0.317
-SynFlow offline		✓	✓	0.672	0.653	0.274
-flow branch				0.643	0.615	0.324

4   Experimental Results

In this section, we present experimental results and analysis on both the single-instance and multiple-instance video object segmentation datasets, as well as our self-constructed synthetic flow data.

4.1   Dataset and Evaluation Metrics

Video Object Segmentation Datasets. We use two high-quality video object segmentation datasets, the DAVIS 2016 [24] dataset and the DAVIS 2017 [25] dataset for single-instance and multiple-instance segmentation tasks, respectively. The DAVIS 2016 [24] dataset is a single-instance video object segmentation dataset that consists of 50 sequences and 3455 annotated frames for real-world moving objects. While the DAVIS 2017 [25] dataset is a larger and more complex video segmentation dataset with multiple targets and instance-level annotations of 10459 frames from 150 videos. We use the pre-defined training set to optimize our framework and its validation set to test the segmentation quality. To evaluate the network performance, we use three measures (evaluation code from [24]): region similarity J, contour accuracy F and temporal stability T.

Synthetic Flow Dataset. For evaluation of the optical flow estimation ability, we test the performance of *SegFlow* on our self-constructed synthetic object flow datasets: DAVIS-sin and DAVIS-multi. DAVIS-sin contains 52000 pairs of frames with a single moving object generated from the train set of the DAVIS 2016 dataset (details in Sect. 2.2), which is split into training (50000) and testing (2000) set. DAVIS-multi contains 52000 pairs of frames with more than one moving object from the train set of the DAVIS 2017 dataset, which is also split into 50000 for training and 2000 for testing. Some examples of these two datasets are shown in Fig. 4. In addition to the traditional end-point-error (EPE) value [9], we also use the segmentation measurement J to evaluate the flow estimation with segmentation mask annotations. Specifically, we warp the segmentation mask of frame t via flow estimation to obtain a flow-based segmentation prediction in

Table 2. Overall segmentation results on the DAVIS 2016 validation set. We analyze various settings for different video object segmentation algorithms, including their performance, training requirement (initial mask, future frames, external data) and add-ons. Methods with the same training settings are listed together.

Method	Initial mask	Future frames	External data	Add-ons	Speed	J mean
OnAVOS [35]	✓		✓	Online update, crf	13 s	0.861
OSVOS [3]	✓		✓	Contour, super-pixel	10 s	0.798
MSK [18]	✓		✓	Optical flow, crf	12 s	0.797
SFL [7]	✓		✓		7.9 s	0.761
MSK-flow [18]	✓		✓		12 s	0.748
CTN [16]	✓		✓	Optical flow	29.95 s	0.735
Lucid [17]	✓			Optical flow, crf	40 s	0.848
Ours+CRF	✓			Crf	11.3 s	0.817
Ours	✓				10.3 s	0.770
Lucid-flow [17]	✓				40 s	0.767
VPN [14]	✓				0.63 s	0.702
PLM [27]	✓				–	0.702
OFL [33]	✓			Optical flow	60 s	0.680
ARP [19]		✓		Candidate regions	–	0.762
LVO [32]		✓		Optical flow	–	0.759
FSEG [13]			✓	Optical flow	7 s	0.707
LMP [31]			✓	Optical flow	18 s	0.700

frame $t + 1$, and then test the region similarity performance J of this warping result. This J value is consistent with the flow prediction quality, as a more accurate flow should provide better temporal correspondence (see Fig. 5 for some visualization examples).

4.2   Ablation Study on Video Object Segmentation

To analyze the necessity and importance of each step in the proposed framework, we carry out extensive ablation studies on the DAVIS 2016 validation set, and summarize the results in Table 1. We compare the performance of *SegFlow* trained with different settings, i.e. without online training, flow branch, synthetic flow in the offline (**SynFlow offline**) and online (**SynFlow online**) training. The detailed settings are explained as follows:

Ours Offline: only using the offline training without the first frame annotation in each test video ('unsupervised' video object segmentation setting).

-flow branch: training the model with only the segmentation branch.

-SynFlow offline: training the network with the SINTEL dataset instead of our synthetic flow data in the offline training process.

-SynFlow online: training the network without synthetic flow (only updating segmentation branch) in the online training process.

As shown in Table 1, the *SegFlow* trained with and without synthetic flow data are compared (i.e. *Ours* vs *−SynFlow online*, *Ours offline* vs *−SynFlow*

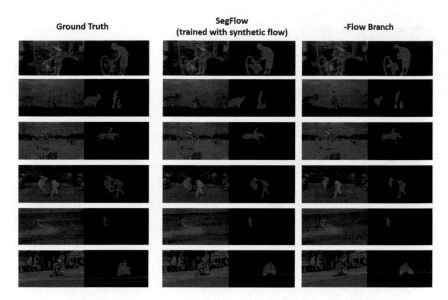

Fig. 6. Segmentation results on the DAVIS 2017 dataset. We show some multi-object segmentation results of *SegFlow* with and without optical flow branch.

Table 3. Segmentation results on the DAVIS 2017 validation set. We show the ablation study and comparison with other algorithms on the DAVIS 2017 dataset. $Ours + CRF$ denotes adding CRF post-processing step; and $-flo$ denotes the network structure without the optical flow branch.

Method	External data	Add-ons	DAVIS 2017
Ours+CRF		Crf	0.524
OSVOS [21]	✓	Contour, super-pixel	0.566
OFL [33]	✓	Optical flow model	0.549
Ours			0.498
Ours-flo			0.475

$Offline$) to validate the effectiveness of the proposed synthetic flow training scheme. The quantitative results in Table 1 show that our synthetic flow data improves *Jmean* by 2.1% in the online training process; and that in the offline training process, the synthetic data can train network as well as external optical flow dataset, demonstrating that our object flow synthesizing training scheme is capable of relieving dataset constraints for joint learning models.

4.3 Segmentation Results

Single-Instance Video Object Segmentation. In Table 2, we compare the performance with state-of-the-art algorithms [3,18,35] and analyze their

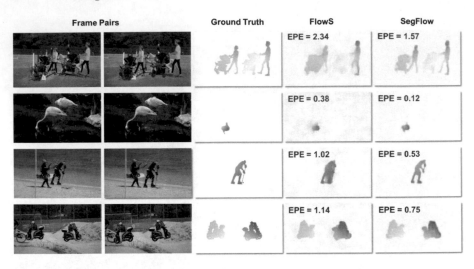

Fig. 7. Examples of flow predictions from the network trained with our synthetic flow.

attributes on the DAVIS 2016 validation set. For semi-supervised methods, our model with synthetic object flow outperforms the original version (SFL) by 5.6%. It also performs favorably against OnAVOS [35], OSVOS [3] and Lucid [17], who require repeatedly updating on test videos (OnAVOS updates network with its own predictions in test videos) or additional inputs (i.e., super-pixels in OSVOS and optical flow in MSK) to achieve higher performance. With image as the only input, the *Jmean* of MSK [18] on the DAVIS validation set is 74.8%, which is lower than ours without post-processing as 77.0%. Besides, with a simple CRF post-processing step, our network performance can achieve the *Jmean* of 81.7%. Among algorithms using only original or synthetic data from the DAVIS dataset, our *SegFlow* outperforms other methods with the same input (Lucid-flow, VPN, PLM).

Multiple-instance Video Object Segmentation. As mentioned above, the synthetic data enables *SegFlow* to learn specific motion patterns in multiple-object segmentation tasks. We test its ability on the DAVIS 2017 validation set. Different from the single-object segmentation task, we change the output channel to the number of instances in test video during online training. This means we still need to finetune one network per video, but the last classification layer changes according to the number of target objects. Table 3 shows the performance on the DAVIS 2017 dataset, which has instance-level segmentation annotations; and Fig. 6 provides some examples of network segmentation predictions with and without flow branch. We once again demonstrate that optical flow plays an important role in video segmentation model *SegFlow*, and that synthetic object flow can help training the flow branch to locate onto a specific target, enabling the joint network to distinguish multiple target instances.

Table 4. Evaluation of flow predictions. *FlowNetS* denotes the results of [9]; *SegFlow* and *Flow* denote the *SegFlow* network and its flow branch, respectively. Both *Flow* and *SegFlow* are trained with our synthetic object flow datasets which have single or multiple moving objects (denoted by *sin* or *mul* in column 3 of the table). Corresponding end-point-errors (EPE) for these two synthetic datasets and the segmentation performance of mask warping results on DAVIS are evaluated for flow quality comparison. Lower EPE and higher J, F denote better performance.

Method	SynFlow	Setting	EPE		DAVIS train		DAVIS test	
			sin	mul	J	F	J	F
FlowNetS [9]			0.945	1.155	0.741	0.725	0.716	0.708
Flow	✓	sin	0.612	0.751	0.749	0.734	0.733	0.714
	✓	mul	0.683	0.742	0.747	0.729	0.736	0.692
SegFlow			0.777	1.007	0.742	0.710	0.726	0.705
	✓	sin	**0.609**	0.731	**0.756**	0.734	**0.744**	0.722
	✓	mul	0.618	**0.721**	0.747	**0.735**	0.733	**0.721**

4.4   Object Flow

To validate the effectiveness of the synthetic object flow, we test the flow predictions on self-constructed synthetic flow datasets (DAVIS-sin and DAVIS-multi), as well as the DAVIS 2016 segmentation dataset. For synthetic flow datasets, we train networks on training samples and test the end-point-error of estimated flow on the validation set. For the video segmentation dataset where optical flow ground truth is not available, we warp the segmentation mask from frame t to frame $t+1$ via flow estimation and evaluate the quality of this warped mask (Fig. 5). As we focus on object area, this metric consistently measures the performance of object flow prediction in real-world videos. Figure 7 shows some example results on synthetic flow datasets, from which we can see the advantage of joint learning (*SegFlow* has better performance than the flow prediction network).

Table 4 shows the results of overall comparison, where *FlowNetS* denotes the model directly obtained from [9], *Flow* denotes the optical flow branch in *SegFlow*, and *SegFlow* denotes the proposed dual-branch network. Networks trained with the synthetic flow have checkmarks in the *SynFlow* column, while *FlowNetS* and *SegFlow* without this checkmark are trained using other optical flow datasets. *sin* and *mul* denotes the training datasets that have single (DAVIS-sin) and multiple (DAVIS-mul) moving objects respectively. The end-point-errors (EPE) are tested on the validation set for both. Results in Table 4 validate that segmentation branch helps boost optical flow estimation (under same condition, *SegFlow* performs consistently better than *Flow*).

In addition, from the evaluation of warped mask on training and validation set of the DAVIS 2016 dataset, we show that the synthetic object flow generated from the same dataset helps boost segmentation better than external optical flow datasets (the last three rows). The best performance in this setting is obtained by *SegFlow* trained with a single moving object, which is 2.8% (0.744 vs 0.716) higher than *FlowNetS* and 1.1% (0.744 vs 0.733) higher than the flow branch on the DAVIS validation set. In Fig. 7, we show some examples of flow predictions and mask warping results.

4.5 Runtime Analysis

SegFlow trained with the synthetic flow data has the same prediction speed as the one trained with external optical flow datasets, i.e. predicting two outputs (segmentation and optical flow) simultaneously at the speed of 0.3 s per frame. When taking the online training step into account, our system runs at 7.9 and 11.3 s per frame for without and with the synthetic flow (averaged over the DAVIS validation set). We present a speed comparison in Table 2.

5 Concluding Remarks

In this paper, we propose a simple and fast object flow synthesizing algorithm which can generate object movement and its corresponding ground truth to relieve the data constraint in joint learning of video object segmentation and optical flow. We use the dual-branch, end-to-end, fully-convolutional network, *SegFlow* as a base net to show that joint prediction models can be trained upon a single video segmentation dataset without manual motion annotations via the proposed object flow synthesizing method. We carry out extensive ablation studies and analysis to validate the effectiveness of the object flow synthesizing training scheme, demonstrating that it has a better training effect than external virtual flow datasets, and that it enables the *SegFlow* to distinguish different targets in multiple-instance video segmentation scenes. Besides, we also propose that this flow synthesizing and joint-training scheme can be easily adapted to other related networks.

References

1. Anderson, R., et al.: Jump: virtual reality video. ACM Trans. Graph. (TOG) **35**(6), 1–13 (2016)
2. Butler, D.J., Wulff, J., Stanley, G.B., Black, M.J.: A naturalistic open source movie for optical flow evaluation. In: Fitzgibbon, A., Lazebnik, S., Perona, P., Sato, Y., Schmid, C. (eds.) ECCV 2012. LNCS, vol. 7577, pp. 611–625. Springer, Heidelberg (2012). https://doi.org/10.1007/978-3-642-33783-3_44
3. Caelles, S., Maninis, K.K., Pont-Tuset, J., Leal-Taixé, L., Cremers, D., Van Gool, L.: One-shot video object segmentation. In: CVPR (2017)
4. Chen, C., Seff, A., Kornhauser, A., Xiao, J.: DeepDriving: learning affordance for direct perception in autonomous driving. In: ICCV (2015)
5. Cheng, H.K., Tai, Y.W., Tang, C.: Modular interactive video object segmentation: interaction-to-mask, propagation and difference-aware fusion. ArXiv abs/2103.07941 (2021)

6. Cheng, J., Tsai, Y.H., Hung, W.C., Wang, S., Yang, M.H.: Fast and accurate online video object segmentation via tracking parts. In: CVPR (2018)
7. Cheng, J., Tsai, Y.H., Wang, S., Yang, M.H.: SegFlow: joint learning for video object segmentation and optical flow. In: ICCV (2017)
8. Cohen, I., Medioni, G.: Detecting and tracking moving objects for video surveillance. In: CVPR (1999)
9. Fischer, P., et al.: FlowNet: learning optical flow with convolutional networks. In: ICCV (2015)
10. Geiger, A., Lenz, P., Urtasun, R.: Are we ready for autonomous driving? The KITTI vision benchmark suite. In: CVPR (2012)
11. He, K., Zhang, X., Ren, S., Sun, J.: Deep residual learning for image recognition. In: CVPR (2016)
12. Ilg, E., Mayer, N., Saikia, T., Keuper, M., Dosovitskiy, A., Brox, T.: FlowNet 2.0: evolution of optical flow estimation with deep networks. In: CVPR (2017)
13. Jain, S.D., Xiong, B., Grauman, K.: FusionSeg: learning to combine motion and appearance for fully automatic segmention of generic objects in videos. arXiv preprint arXiv:1701.05384 (2017)
14. Jampani, V., Gadde, R., Gehler, P.V.: Video propagation networks. arXiv preprint arXiv:1612.05478 (2016)
15. Jampani, V., Gadde, R., Gehler, P.V.: Video propagation networks. In: CVPR (2017)
16. Jang, W.D., Kim, C.S.: Online video object segmentation via convolutional trident network. In: CVPR (2017)
17. Khoreva, A., Benenson, R., Ilg, E., Brox, T., Schiele, B.: Lucid data dreaming for object tracking. arXiv:1703.09554 (2017)
18. Khoreva, A., Perazzi, F., Benenson, R., Schiele, B., Sorkine-Hornung, A.: Learning video object segmentation from static images. In: CVPR (2017)
19. Koh, Y.J., Kim, C.S.: Primary object segmentation in videos based on region augmentation and reduction. In: CVPR (2017)
20. Long, J., Shelhamer, E., Darrell, T.: Fully convolutional networks for semantic segmentation. In: CVPR (2015)
21. Maninis, K.K., et al.: Video object segmentation without temporal information. TPAMI 41, 1515–1530 (2018)
22. Mayer, N., et al.: A large dataset to train convolutional networks for disparity, optical flow, and scene flow estimation. In: CVPR (2016)
23. Oh, S., Lee, J.Y., Xu, N., Kim, S.: Video object segmentation using space-time memory networks. In: 2019 IEEE/CVF International Conference on Computer Vision (ICCV), pp. 9225–9234 (2019)
24. Perazzi, F., Pont-Tuset, J., McWilliams, B., Gool, L.V., Gross, M., Sorkine-Hornung, A.: A benchmark dataset and evaluation methodology for video object segmentation. In: CVPR (2016)
25. Pont-Tuset, J., Perazzi, F., Caelles, S., Arbeláez, P., Sorkine-Hornung, A., Van Gool, L.: The 2017 DAVIS challenge on video object segmentation. arXiv:1704.00675 (2017)
26. Ros, G., Ramos, S., Granados, M., Bakhtiary, A., Vazquez, D., Lopez, A.M.: Vision-based offline-online perception paradigm for autonomous driving. In: WACV (2015)
27. Shin Yoon, J., Rameau, F., Kim, J., Lee, S., Shin, S., So Kweon, I.: Pixel-level matching for video object segmentation using convolutional neural networks. In: ICCV (2017)
28. Srivastava, R.K., Greff, K., Schmidhuber, J.: Highway networks. In: ICML (2015)

29. Sun, C., Lu, H.: Interactive video segmentation via local appearance model. TCSVT **27**, 1491–1501 (2016)
30. Tian, Y.L., Lu, M., Hampapur, A.: Robust and efficient foreground analysis for real-time video surveillance. In: CVPR (2005)
31. Tokmakov, P., Alahari, K., Schmid, C.: Learning motion patterns in videos. In: CVPR (2017)
32. Tokmakov, P., Alahari, K., Schmid, C.: Learning video object segmentation with visual memory. In: ICCV (2017)
33. Tsai, Y.H., Yang, M.H., Black, M.J.: Video segmentation via object flow. In: CVPR (2016)
34. Voigtlaender, P., Leibe, B.: Online adaptation of convolutional neural networks for the 2017 DAVIS challenge on video object segmentation. In: CVPR Workshop (2017)
35. Voigtlaender, P., Leibe, B.: Online adaptation of convolutional neural networks for video object segmentation. In: BMVC (2017)
36. Wug Oh, S., Lee, J.Y., Sunkavalli, K., Joo Kim, S.: Fast video object segmentation by reference-guided mask propagation. In: CVPR (2018)
37. Yang, Z., Wei, Y., Yang, Y.: Collaborative video object segmentation by foreground-background integration. In: Vedaldi, A., Bischof, H., Brox, T., Frahm, J.-M. (eds.) ECCV 2020. LNCS, vol. 12350, pp. 332–348. Springer, Cham (2020). https://doi.org/10.1007/978-3-030-58558-7_20
38. Zhou, T., Wang, S., Zhou, Y., Yao, Y., Li, J., Shao, L.: Motion-attentive transition for zero-shot video object segmentation. In: AAAI (2020)

Recognition of Bending Deformed Pipe Sections in Geological Disaster Area Based on an Ensemble Learning Model

Zhao Ziqi[1], Chen Chao[1], Dai Jinyang[1], Liu Shen[1], Li Bo[2,3], and Liu Xiaoben[1](✉)

[1] National Engineering Laboratory for Pipeline Safety, MOE Key Laboratory of Petroleum Engineering, Beijing Key Laboratory of Urban Oil and Gas Distribution Technology, China University of Petroleum, Beijing 102249, China
xiaobenliu@cup.edu.cn
[2] Shenyang Longchang Pipeline Inspection Co., Ltd., Shenyang 11031, China
[3] Changping, Beijing 102249, China

Abstract. At present, the artificial recognition method is mainly used to identify IMU strain detection data of the whole pipe section by segment, which has some problems such as low efficiency, high cost and long cycle. Therefore, this paper realizes the intelligent recognition of the Bending Deformed Section in Geological Disaster Area (BDPIGDA) by establishing an ensemble learning model. Firstly, it is statistically obtained that the pipe sections with bending strain value exceeding 0.125% in an oil pipe include bend, dent section, BDPIGDA. Then, combined with geometric detection data, sample data of different pipe sections are intercepted, and 11 typical data feature values are extracted. Through principal component analysis, kernel principal component analysis, and independent component analysis, the data dimension of the 11 feature data is reduced. Finally, an ensemble learning model combining support vector machine and K-means clustering is established. The research results show that the accuracy rate of the test set of the model is 93.26%, and the recognition rate of the bent deformed pipe section in the geological disaster area is 88.70%, which meets the engineering requirements and provides a certain reference for pipe integrity management.

Keywords: Data dimension reduction · Support vector machine · K-means clustering · Ensemble learning

1 Introduction

Reasonable and effective analysis of pipe parameters is an important factor to ensure the safe operation [1]. The strain of pipes in abnormal state and specific areas would have certain change characteristics. The strain fluctuation in bend and dent sections is relatively large, and the strain fluctuation range in bending deformed pipe sections is relatively wide [2].

At present, IMU detection technology is an effective method to quickly obtain the bending strain of the whole pipe. The gyroscope is used to obtain the rotational angular velocity of the object in three directions, and the accelerometer is used to obtain

© Springer Nature Singapore Pte Ltd. 2021
Y. Wang and W. Song (Eds.): IGTA 2021, CCIS 1480, pp. 125–131, 2021.
https://doi.org/10.1007/978-981-16-7189-0_10

the motion acceleration of the object in three directions, which can obtain the speed, position and attitude information of the detector at any time, and then calculate and transform the horizontal strain and vertical strain of the pipe [3–5]. According to the requirements of Appendix I of GB32167-2015 "Code for Integrity Management of Oil and Gas Transmission Pipes", the performance specifications measured by IMU shall meet the bending characteristics of pipes with bending deformed curvature $> 1/400$ D (strain value exceeds 0.125%) identified by single test. It is an important measure to ensure safe operation of pipes on screening out dangerous sections of bending deformed from these characteristics.

By analyzing the strain data obtained after transformation, Zhao Xiaoming, et al. [6] judged high-risk points with potential threats combined with geometric detection data based on artificial recognition to locate pipe defects. Fang Weilun, et al. [7] summarized the strain data characteristics of bend, dent and BDPIGDA based on IMU internal detection data. However, for massive strain detection data in IMU, artificial recognition takes a long time, and the recognition results vary from person to person.

In this paper, a machine learning based recognition method based on ensemble learning model is proposed by taking IMU strain detection data B and C of an oil pipe in 2019 as sample data, in which B is the training set and C is the test set. Firstly, the strain data of pipe section is preprocessed, According to the physical characteristics of the pipe section, three dimensionality reduction methods, principal component analysis (PCA), kernel principal component analysis (KPCA) and independent component analysis (ICA), are used to reduce the dimensions of 11 characteristics of the sample data. Then, support vector machine and K-means clustering [8, 9] are used to build a classification method. Finally, an ensemble learning model combining the two classification methods is proposed to realize an intelligent classification of bending deformed segments.

2   Feature Engineering

2.1   Typical Pipe Section

Combined with geometric detection data labels and research by Fang Weilun scholars, typical pipe sections in bend, dent and Bending Deformed Pipe Section in Geological Disaster Area (BDPIGDA) are shown in Fig. 1. The strain characteristics of bends are often long, thin and spiky, and the vertical strain mostly exceeds 0.5%. The Dent is short, thick and sharp, and the strain value is relatively flat. The BDPIGDA based on IMU data are mainly divided into "W" shape.

2.2   IMU Data Feature Construction

According to the physical meaning of IMU strain data and referring to the relevant methods of signal analysis [10, 11], this paper extracts 11 eigenvalues from each sample data, namely length, amplitude, peak-to-peak value, minimum value, mean value, standard deviation, skewness, kurtosis, peak factor, pulse factor and margin factor.

PCA, KPCA and ICA are used to reduce the dimension of data. Among them, PCA is a linear feature extraction method, which recombines P observed variables into M new

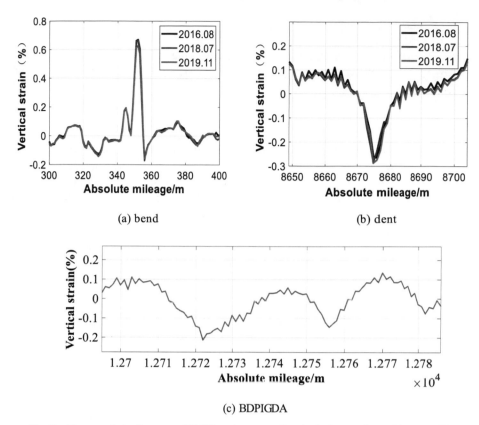

(a) bend   (b) dent

(c) BDPIGDA

Fig. 1. Characteristic diagram of IMU strain curve of typical pipe section with anomalies

unrelated variables through orthogonal transformation, and the transformed variables are called main components. In this paper, the first three principal components are selected, and the cumulative contribution rate is 87.82%. KPCA introduces kernel function, thus mapping nonlinear original data to high-dimensional space or even infinite-dimensional space, making it linearly separable. Then, through PCA dimension reduction, this paper selects the first three kernel principal components, with a cumulative contribution rate of 91.99%. ICA is a dimension reduction method to find potential independent factors, which can effectively unmix data [12]. After subsequent model tests, it is found that selecting the first two factors has the best recognition effect.

3 Ensemble Learning

3.1 Support Vector Machine

The basic idea of Support Vector Machine (SVM) is that nonlinear inseparable problems can be partitioned by hyperplane. The specific method is to map the low-dimensional feature space where the sample data is located to the high-dimensional space first, which completes the construction of the classification model by high-dimensional support plane division [13].

Because the support vector machine model adopts the optimization idea of minimizing structural risk, the generalization ability of the model often needs to be further tested after training.

3.2 K-means Clustering

K-means clustering is a typical unsupervised learning method based on distance measurement to judge the distribution category of samples in feature space. The process of this method is shown in Fig. 2. Commonly used distance measures include Euclidean distance, correlation distance, Manhattan distance, and cosine similarity [14, 15].

At the beginning of the algorithm, K points are randomly selected as the initial clustering center points in the sample space. The distance from the sample point to the center point is calculated, then each sample point is classified according to the nearest distance. The sample points divided into the same category are solved with the average value, and finally the average value is used to replace the initial clustering center, which iterates repeatedly until the clustering center does not change or the number of iterations reaches the upper limit.

This clustering method is not applicable to some interlaced feature spaces, so it is necessary to use generalization test sets to further test the performance of the model.

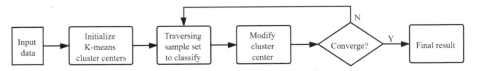

Fig. 2. K-means clustering model diagram

3.3 Ensemble Learning Based on Voting Method

First of all, this paper selects 1050 segments of IMU strain data from the research data (i.e. 350 segments of bend, dent and BDPIGDA) as training sets, and trains the SVM model and the K-means model respectively. The recognition rate of the two models in the three feature spaces is over 80%, and they have good classification performance.

The voting method is adopted to ensemble the learning model [16], and the support vector machine and K-means clustering are combined to improve the accuracy and robustness of the model. Figure 3 is a specific ensemble learning flow chart. Firstly, according to the requirements of national standard, the pipe section with strain amplitude exceeding 0.125% and length exceeding 12 m is screened out by the built-in function Findpeaks of Matlab. The bends and dents are marked according to the labels of geometric detection data. Next, six classifiers are fused for ensemble learning by voting method. Finally, bends and dent in the ensemble learning recognition results are checked according to the geometric test results, and the recognition results of geological disaster segments are output.

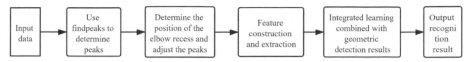

Fig. 3. Ensemble learning flow chart

3.4 Analysis of Results

The ensemble learning model passed the Five-fold cross-validation on the training set, and the model recognition rate was 95.43%. On the test set, the model recognition rate is 93.26%, as shown in Fig. 4, which is slightly lower than the training set.

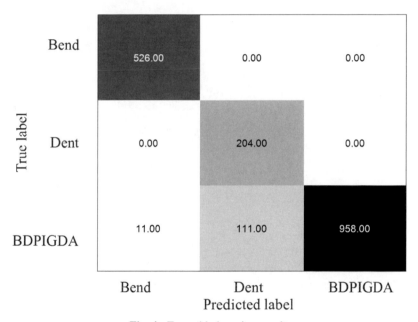

Fig. 4. Ensemble learning results

From the test set recognition results of each classifier (Table 1), it can be seen that among the three feature spaces of the test set, SVM and K-means methods have the highest recognition rate in ICA feature space, both exceeding 80%. Compared with a single classifier, the ensemble learning model has the best generalization ability, especially the recognition rate of BDPIGDA is the highest, reaching 88.70%.

As shown in Fig. 5, a total of 699 pipe sections with large strain were found in the ensemble learning model, including 40 suspected bends, 259 suspected dents and 400 suspected BDPIGDA. According to the manual check after drawing, it is determined that the maximum strain amplitude in these pipe sections exceeds 0.125%, and the strain curves of 40 suspected bends and 40 suspected Dents show that the model has better recognition ability.

Table 1. Comparison of test set recognition results of each classifier

Category	Recognition rate			
	Bend	Dent	BDPIGDA	Overall
ICA + SVM	98.78%	56.76%	88.50%	87.31%
PCA + SVM	90.19%	87.93%	42.71%	73.61%
KPCA + SVM	95.58%	50.17%	49.14%	65.07%
ICA + K-means	98.00%	74.29%	70.57%	80.95%
PCA + K-means	94.86%	49.14%	72.86%	72.29%
KPCA + K-means	98.57%	58.86%	76.00%	77.81%
Ensemble learning	100%	100%	88.70%	93.26%

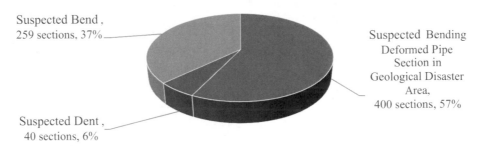

Suspected Bend, 259 sections, 37%

Suspected Bending Deformed Pipe Section in Geological Disaster Area, 400 sections, 57%

Suspected Dent, 40 sections, 6%

Fig. 5. Newly discovered pipe section predicted by the model

4   Conclusion

The bending strain data of the whole pipe are obtained by IMU test, and the following conclusions can be drawn by analyzing the data by using the ensemble learning model based on voting method:

(1) In this paper, three dimension reduction methods, PCA, KPCA and ICA, are used to reduce the dimension of 11-dimensional feature data. ICA realizes data unmixing by finding independent components. After subsequent tests, ICA is found to be the best dimension reduction method among the three dimension reduction methods.

(2) In the test set of SVM and K-means clustering methods in these three feature spaces, only in ICA feature space, the recognition rate exceeds 80%, indicating that ICA + SVM and ICA + K-means have better generalization performance.

(3) Comparatively speaking, the generalization ability of ensemble learning is stronger than that of a single model. The generalization accuracy of the BDPIGDA is close to 90%, which can meet the engineering requirements, reduce the cost of artificial recognition, and provide a certain reference for pipe integrity management.

Acknowledgements. This research has been co-financed by National Science Foundation of China (Grant No. 52004314), Beijing Municipal Natural Science Foundation (Grant No. 8214053), Tianshan Youth Program (Grant No. 2019Q088), the Open Project Program of Beijing Key Laboratory of Pipeline Critical Technology and Equipment for Deepwater Oil & Gas Development (Grant No. BIPT2020005), Science Foundation of China University of Petroleum, Beijing (No. 2462018YJRC019, No. 2462020YXZZ045).

References

1. Dong, S.: Review and development suggestions of oil and gas pipe integrity management in China in the past 20 years. Oil Gas Storage Transp. **39**(03), 241–261 (2020)
2. Jeremie, J.C., Sylvain, C., Mohamed, E., et al.: Understanding pipe strain conditions: case studies between ILI axial and ILI bending measurement techniques. In: Calgary: International Pipe Conference, IPC2018-78577 (2018)
3. Tan, D., Zheng, J., Ma, Y., et al.: Displacement monitoring of Mohe-Daqing oil pipeline in the permafrost area. Oil Gas Storage Transp. **31**(10), 737–739 (2012)
4. Li, S.: Current situation of oil/gas pipe monitoring technologies. Oil Gas Storage Transp. **33**(2), 129–134 (2014)
5. Bao, Q., Shuai, J.: Research progress on internal detection technology of oil and gas pipes. Contemp. Chem. Ind. **46**(2), 298–301 (2017)
6. Zhao, X., et al.: Identification and evaluation on bending deformation of China-Russia Eastern gas pipeline. Oil Gas Storage Transp. **39**(07), 763–768 (2020)
7. Fang, W., Liu, X., Zhang, H., Chen, P., Li, R., Zhao, X.: Study on recognition method of pipe thaw settlement risk section in frozen soil region based on IMU detection data. Oil Gas Storage Transp. (2020)
8. Hu, T., Guo, Q., Sun, H.: Electricity theft test based on stack decorrelation self-encoder and support vector machine. Power Syst. Autom. **43**(647(01)), 162–170 (2019)
9. Liu, X., Zhang, H., Xia, M., Liang, L., Zheng, W., Li, M.: Pipeline leakage recognition based on principal component analysis and neural network. Oil Gas Storage Transp. **34**(07), 737–740 (2015)
10. Xiang, J., Yong, Z., Wang, D.: A method and device for processing abnormal data of fan based on quartile box diagram. CN106897941A (2017)
11. Liu, X., Zhang, H., Xia, M., Liang, L., Zheng, W., Li, M.: Pipe leakage recognition method based on principal component analysis and neural network. Oil Gas Storage Transp. **34**(07), 737–740 (2015)
12. Xia, M., Liu, X., Chen, Y., et al.: Diesel engine fault diagnosis method based on principal component analysis and self-organizing neural network. Comput. Appl. **34**(2), 184–185, 229 (2014)
13. Mita, J.H., Babu, C.G., Shankar, M.G.: Performance analysis of dimensionality reduction using PCA, KPCA and LLE for ECG signals. IOP Conf. Ser. Mater. Sci. Eng. **1084**(1), 012005 (8 pp.) (2021)
14. Lan, L., Wang, Z., et al.: Scaling up kernel SVM on limited resources: a low-rank linearization approach. IEEE Trans. Neural Netw. Learn. Syst. **30**(2), 369–378 (2019)
15. Zheng, H., Jiang, H., Zhang, X.: Application of poly KPCA in fault diagnosis of high-dimensional bearings. Mach. Tool Press (11) (2020)
16. Wu, J., Li, Q., Zhao, J., et al.: Face recognition based on improved ICA and RBF neural network. J. Xi'an Univ. Posts Telecommun. **23**(134(05)), 22–26 (2018)

Memory Bank Clustering for Self-supervised Contrastive Learning

Yiqing Hao[1,2], Gaoyun An[1,2($\boxtimes$)], and Qiuqi Ruan[1,2]

[1] Institute of Information Science, Beijing Jiaotong University, Beijing 100044, China
{19140011,gyan,qqruan}@bjtu.edu.cn
[2] Beijing Key Laboratory of Advanced Information Science and Network
Technology, Beijing 100044, China

Abstract. Contrastive Learning aims at embedding positive samples close to each other and push away features from negative samples. This paper analyzed different contrastive learning architectures based on the memory bank network. The existing memory-bank-based model can only store global features across few data batches due to the limited memory bank size, and updating these features can cause the feature drift problem. After analyzing these issues above, a network for contrastive learning with visual representations is proposed in this paper. First, the model is combined with a memory bank and memory feature clustering mechanism; Second, a new feature clustering method is proposed for memory bank network to find and store cross-epoch global feature centers for training epochs based on the memory bank architecture. Third, the centers in memory bank are treated as class features to construct positive and negative samples with current batch data and apply contrastive learning methods to optimize a feature encoder to learn a better feature representation. Finally, this paper designed a training pipeline to update the memory bank and encoder individually to circumvent the feature drift problem. To test the performance of proposed memory bank clustering method with on unsupervised image classification, our experiment used a self-supervised online evaluator with an extra non-linear layer. The experiment results show that our proposed model can achieve good performance on image classification tasks.

Keywords: Contrastive learning · Self-supervised learning · Image classification

1 Introduction

Deep learning methods can learn patterns from the huge amount of data. With the continuous development of deep learning, in most Computer Vision (CV) tasks, the supervised learning methods can learn feature representation from large amounts of labeled samples. The supervised learning methods may have problems with generalization caused by model overfitting or require a large amount of human-labeled data. Under these circumstances, unsupervised and self-supervised methods gradually emerge, while these

Y. Hao—Supported by the National Natural Science Foundation of China (62072028 and 61772067).

methods do not need labeled data and can achieve similar performance to supervised algorithms.

Learning a better feature representation is essential for unsupervised or self-supervised algorithms. Unsupervised deep learning methods map the input to high-dimensional latent space and apply clustering algorithms to learn feature representation from the input data.

Self-supervised contrastive learning methods can learn feature representation by similarity function that measures how similar or related two feature representations are. Contrastive Learning is a discriminative approach, which often uses similarity measurement methods to divide the positive and negative samples from input data, then group the distribution of positive feature embeddings closer and make negative feature embeddings farther away.

In image classification tasks, self-supervised contrastive learning methods need to know the similar sample or diverse samples and use a suitable loss function to optimize the deep neural encoder network to group the feature of similar samples closer and farther diverse samples. Metric Learning methods aim to establish similarity or dissimilarity between objects. Contrastive Learning and Metric Learning are closely related. Matric learning methods can measure the distance between features in high-dimensional space and distinguish positive and negative samples from the features. However, it is often challenging to design metrics suited to the detailed data and the task. There are different deep metric learning algorithms in different scenarios. The objective function is constructed for deep contrastive algorithms, and the cost function (loss) is minimized to get better feature representations.

This paper compares two commonly used architectures for contrastive Learning, the end-to-end architecture, and Memory Bank based architecture in Sect. 2. Our main work is: (1) This paper proposes a contrastive learning network architecture with a dynamically updated memory bank for image classification. (2) A feature clustering method is designed to store and update the feature center with a memory bank. (3) A contrastive loss is applied to the model. The loss based on feature centers stored in the memory bank to optimize our feature encoder. We tested the results of the proposed model in this paper on multiple datasets. The experiments show that our proposed model has good performance on self-supervised image classification tasks. The center features stored in the memory bank can provide global information for optimization so the model can get a good feature representation.

2   Related Work

As a self-supervised method, contrastive Learning relies on the number of samples for generating feature representations then learning the differences between similar features and different types of features using metric methods that measures how similar or related two objects are. Memory bank can store feature representations for accessing negative samples [3]. This paper analyzes some memory bank related contrastive learning methods. The architecture of different architecture pipelines for contrastive learning, as Fig. 1, end-to-end contrastive training architecture [14, 23], consists of two encoders generating feature representation for positive samples and treat others as the negative. The memory bank based contrastive architecture contains memory bank to store and retrieve feature embeddings of negative and positive samples [12, 20].

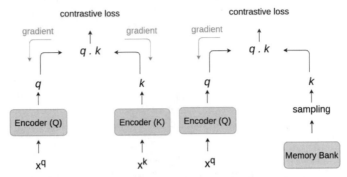

Fig. 1. Different architecture for Contrastive Learning [14]: End-to-End contrastive learning architecture (left), Memory bank based architecture (right).

2.1 Self Supervised Contrastive Learning

We have studied earlier works in the field of contrastive Learning with self-supervised learning [4, 5, 15]. The related methods include constructing positive and negative samples, then grouping positive samples closer while differing negative samples in the feature representation from each other.

To construct positive and negative sample pairs, some methods [6, 7, 12] use data augmentation methods. Most commonly used data augmentation operators can be divided into the following parts: color transformation, geometric transformation, context-based transformation, and cross-modal-based transformation. These data augmentation methods are applied to construct positive and negative sample pairs. An example of augmented data is shown in Fig. 2.

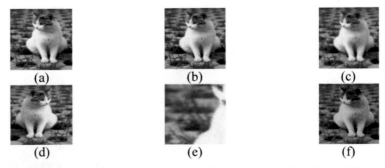

Fig. 2. Illustrations of augmentation operators used in our training data. Image (a) is the original image. From left to right and top to bottom, the operators include (b) Color Distort (drop), (c) Gaussian Blur, (d) Random Flip, (e) Random Crop and (f) Color Distort.

2.2 Memory Bank Based Method

Memory network [21] provides a memory component that can be read and written to, which has long-term storage capabilities. As one of the memory network structures, Memory Bank has been widely used in contrastive learning methods [18, 20, 22]. In theory, the more feature information the model can be referred to, the better the contrastive model will be. It is important to maintain separate feature embeddings and accumulate many feature representations used as positive and negative samples during training.

The feature embeddings stored in the memory bank may become outdated as the model optimization process continuously updates the encoder network. This phenomenon is called feature drift. The feature drift of input x at t-th iteration with step Δt is defined [20] as:

$$D(x, t; \Delta t) := \left\| f\left(x; \theta^t\right) - f\left(x; \theta^{t-\Delta t}\right) \right\| \tag{1}$$

To avoid feature drift problem caused by the encoder update, previous work [20] apply a pre-train method, other approaches such as MOCO [12] use momentum to update the encoder slowly, while this paper propose a new train method to avoid feature drift, which will be discussed at Sect. 3.

The work of this paper is based on the contrastive learning methods with memory bank and is improved based on the existing method. First, this paper proposes a contrastive learning network architecture with a dynamically updated memory bank for image classification, compared to general ideas with memory bank, our method can store more features in memory bank for feature representation learning. Second, we design a memory bank clustering method to store and update the feature centers based on the memory bank. The model can construct positive and negative pairs from the feature representation obtained after memory bank clustering. Third, a contrastive loss is designed based on centers stored in the memory bank to optimize our feature encoder.

3 Method

3.1 Contrastive Learning with Memory Bank Clustering

This section will introduce our model architecture shown in Fig. 3, the contrastive loss based on memory bank clustering, and the network training pipeline. First, the experiments use random data augmentation operator to construct samples in the training data batch. Second, a dynamically updated memory bank architecture is used to store the class center from feature embeddings and use the metric learning algorithm to distinguish positive and negative samples corresponding to each feature, then use a memory-feature-clustering based loss function to optimize the encoder network, the above step called as Memory Bank Clustering. The contrastive learning method can be used to update the encoder network M Third, and the train pipeline update the feature encoder (M) and memory bank (B) separately to reduce the error of feature drift [20].

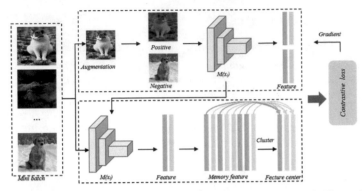

Fig. 3. The architecture of our self-supervised Contrastive learning model. The training of the model apply the data augmentation method for every input data batch and use a deep convolutional neural network M as the backbone encoder to extract feature F. The feature F will be stored in a dynamic memory bank and cluster into feature center C, the model use a batch feature F based contrastive loss L_b and a memory center based C contrastive loss L_{mem} to optimize our encoder network M.

3.2 Training with Dynamic Memory Bank

As Fig. 3 shows, the training apply an end-to-end contrastive learning method with data augmentation to an encoder network [13]. The model use the origin training data and augmented data forward to the encoder and get two corresponding features as a similar pair.

Our data augmentation method is inspired by SimCLR [14]. Given an input image I, the model use several different data augmentation methods to generate positive input pair $\tilde{I}$, the illustrations of data augmentation operators shown in Fig. 2. The method combines I and $\tilde{I}$ together as the output feature of the encoder. During the training phase of the model, then randomly select one data augmentation operator for each input image I_i and output one augmented image $\tilde{I}_i$. So our encoder input data batch B of the model consists of:

$$B = \left\{ \left(I_1, \tilde{I_1} \right), \left(I_2, \tilde{I_2} \right) \cdots \left(I_{bs}, \tilde{I_{bs}} \right) \right\}. \tag{2}$$

Where bs denotes training batch size. Each image pair $\left(I_i, \tilde{I_i} \right)$ as a positive sample pair and $\left(I_j, \tilde{I_k} \right)$, where $j \neq k$ as a negative pair. Then use batch data based contrastive loss L_b to optimize the encoder M. Given an input batch B, and encoder network by M, The output features show as:

$$\left(F_i, \tilde{F}_i \right) = M \left(I_i, \tilde{I_i} \right). \tag{3}$$

For memory update, F_i denotes the feature of input I_i extract by M where $F_i \in R^{(1 \times c)}$. In order to update our encoder network, $(F_i, \tilde{F}_i)$ as a positive pair and others as negative pair. For every step of our training process, it will store the feature F to the memory bank and apply the clustering method to update the feature center in memory bank. The model use the similarity measurement method $sim(x, y)$ to calculate the similarity between center C and feature F_i, if similarity between F_i and memory center C_i greater than similarity threshold T^*, then use all features. $(F_m, \cdots, F_n)$ similar to center C_i to update C_i as a new center C_{new}:

$$C_{new} = \sum_i^m \left(\frac{sim(F_i, C_i) \times D(F_i, C_i)}{\sum_i^m (sim(F_i, C_i))} \right) + C_i \qquad (4)$$

Where $sim(F_i, C_i)$ is the similarity value between feature F_i and center C_i, m is the total number of features stored in the memory bank, F include features where similarity $sim(F_i, C_i)$ between center C_i greater than similarity threshold T^*, D represents the direction measurement from the feature to the corresponding center. In our experiment, our experiments use cosine similarity as $sim(x_1, x_2)$

$$sim(x_1, x_2) = \frac{x_1 \cdot x_2}{max(\|x_1\|_2 \cdot \|x_2\|_2, \varepsilon)} \qquad (5)$$

Two memory bank merge methods are designed in our training phase of the memory center dynamic merge stage which called memory bank clustering. The memory update and merge methods in SimCLR [6] and MoCo [7] are shown in Fig. 4. We use the pre-trained encoder M to generate a batch of features F. For every step in the training epoch, we add the feature F to the memory bank and use similarity matrix m_f and threshold T^* to update the feature center. If one feature F_i is not similar to other centers $(sim(F_i, C) \leq T^*)$, we consider it a center C_m. After updating the memory bank with all batch data in this training step, we merge our memory bank center use $sim(C_i, C_j)$ where $i, j \in M$ and T^*, same as updating the memory center with the feature. After the update

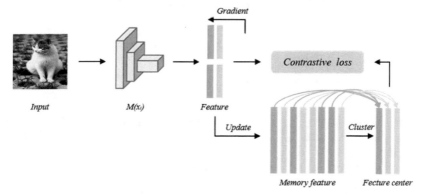

Fig. 4. The train pipeline with memory bank. We apply the data augmentation method for every input data batch and use a deep convolutional neural network M as backbone encoder to extract feature F. The feature F will merge with the memory centers in every train step and we use batch loss L_b and memory-bank based loss L_{mem} to update encoder parameters and memory feature in each training step.

and merge step, the training of the model use memory loss L_{mem} and batch-based loss L_b optimizer our encoder. When every train epoch end, memory bank will be emptied to reduce the feature drift error, so the model only calculate the center from the features of the current epoch to avoid the feature drift problem.

In this paper, another train pipeline is designed for memory bank updating and clustering. As mentioned before, if updating the encoder and memory bank simultaneously, the feature drift will affect our network performance. Due to optimizing our encoder network M in one train step, the model get a feature F_C with data I, and in the next train epoch, if using the same data I, encoder will output a feature F_n, there will be differences between F_c and F_n. The training of the model try to update the memory bank with the encoder network M throughout the epoch and the parameters of the encoder network are frozen. Our training process is as Fig. 5. The training pipeline with memory bank and memory bank clustering method as designed. M denotes the encoder network, F denotes the feature, and C denotes the center in the memory bank. After pretraining M, freeze M and use an epoch to get feature F to update and merge with C. In this epoch, the model only calculate batch loss as experiment data and do not apply backward. Next epoch, unfreeze M, freeze the memory bank and use batch data based loss L_b and memory loss L_{mem} optimize M. After a memory training and encoder network training epoch, the features stored are cleared in memory bank so the feature drift in stored features can be minimized.

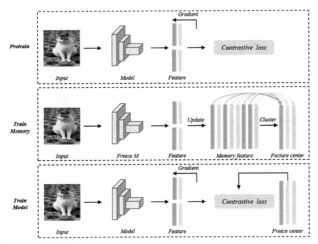

Fig. 5. The pipeline of our model. After pretraining of M, we freeze M and only use the training data forward to M then training our memory bank and apply clustering method to the feature stored in the memory bank. Next epoch, we unfreeze M and freeze our memory bank, use batch loss L_b and memory loss L_{mem} optimize M.

3.3   Contrastive Loss with Dynamic Memory Bank

Contrastive loss was designed to encourage positive pairs to be as close as possible and negative pairs to be apart over a given threshold [10]. The loss function L consists of two parts, batch sample based contrastive loss L_b and memory bank based contrastive loss L_{mem}.

$$L = L_b + L_{mem} \tag{6}$$

The model applies cosine similarity as measurement method of positive and negative samples. The batch based loss L_b as:

$$L_b = -[log(exp(sim(F_i, \widetilde{F_i}))) - log(\sum_{k \neq j} exp(sim(F_k, F_j)))] \tag{7}$$

For each center in the memory bank, the memory bank based contrastive loss should be able to make the feature and center encoding of the same class close, while far away from other centers. In order to achieve our goal, the model use the same cosine similarity to find the memory bank centers $\{C_m, \cdots, C_n\}$ corresponding to each feature F_i as a positive pair. The model use similarity matrix result S_{label} for each feature F_i with every center in memory bank as a softmax label and apply label smooth to S_{label}, the memory loss as:

$$L_{mem} = mean(\sum_F log\left(\frac{\exp(S_{label})_i}{\sum_j exp(S_{label})_j}\right)) \tag{8}$$

Where i denotes the max similarity between memory bank centers and current feature F_i. j denotes the number of centers in memory bank.

As mentioned before, Due to the feature drift problem mentioned before as Eq. 1, if we update the feature encoder M and memory bank center simultaneously, the feature drift will affect our model performance. We test two update methods, update encoder and memory bank at the same training step or use one epoch train encoder with a batch loss L_b and use one epoch generate feature and merge them to center and store in the memory bank. We try both of the training methods with batch based loss L_b and memory center-based loss L_{mem} with the same training hyperparameters.

We also try ArcFace [8] loss and Dsam [16] loss as our memory feature based loss. ArcFace loss [8] use margin to measure the distance between different classes. We calculate a similarity matrix between F and C and use similarity threshold T^* to divide the memory bank center as a class label for each feature F_i. We consider each $Sim(F_i, C_j)$ as loss weight in ArcFace loss. s denotes the similarity weight for each feature F_i.

$$L = -\frac{1}{N} \sum_{i=1}^{N} log \frac{e^{s \cdot (cos(\theta_{C_i} + m))}}{e^{s \cdot (cos(\theta_{C_i} + m))} + \sum_{j=1, j \neq C_i}^{n} e^{s \cdot cos\theta_j}} \tag{9}$$

4   Experiment

To verify our proposed method, we use pytorch-lightning [9] to build self-supervised contrastive network as Fig. 3, use Resnet [13] as feature encoder M, The model is trained and evaluated on Cifar10 [17], Stl-10 [26] and Imagenet-10 [25] dataset. When evaluating our model, our experiment follows the linear evaluation protocol [1, 24].

4.1 Experiment Details

During the training phase of our network, the encoder network is trained n epoch only with the batch sample based loss L_b. This training method [20] can minimize the feature drift. In our experiment, pretrain the model for 4000 steps, about 120 epochs only with L_b, then use both memory loss L_{mem} and batch data based loss L_b to optimize the encoder network. Larger batch size can usually make contrastive learning model obtain better results. If the training data set contains multiple categories, the feature saved in memory before clustering will increase and may occupy a lot of GPU memory, so the experiments only test the model on datasets of fewer categories.

To find the best parameters for the training of our model, we use hydra [27] to configure the parameters and use auto tune for each dataset. To find the best settings, the experiments set different hyperparameters like Table 1.

Table 1. The hyperparameters. We use hydra to tune the hyperparameters for each training phase.

Parameter	Min value	Max value	Next training value
τ	0.1	0.4	$0.05 + \tau$
L	0.000001	0.001	$10 \times$ L
T^*	0.4	0.9	$0.1 + T^*$
i	20	120	$10 + i$
Batch size	32	256	$16 +$ B

In the experiments, due to GPU memory limitations, there is a trade-off between larger batch size and dynamic update memory bank. The feature saved in memory bank will the stored feature information will take up a lot of GPU memory space. At the start of training, if the experiment trained the memory bank with a high clustering threshold, the GPU often runs out of memory because the memory bank stored to many features as data category centers. The final experiment hyperparameters shows in Table 2.

When finding the hyperparameters, we find that the model always get benefit from larger batch size. But the GPU memory limit the batch size, and if we chose a lower merge threshold T^*, feature in memory bank can be merged well but the clustering result for centers may have a bad influence on contrastive method for learning feature representation.

Dynamic threshold T^* is also applied in the experiments. If we fix the value of T^* at the start of training epoch, the similarity of same class features always less than T^*. And we try to use dynamic T^*. We set the T^* a small value, as the training of our model, keep increase the value of T^* to a max value. This have the same effect as our pretrain pre-training strategy and there is almost no difference in the results of the two methods. It is necessary to find a find a suitable threshold is really import impact on the performance of the model.

Table 2. Hyperparameters for each dataset in our final experiments

Hyperparameters	Cifar 10	Stl10	Imagenet-10
τ	0.1	0.1	0.3
L	0.0001	0.0001	0.0001
T^*	0.7	0.6	0.7
Batch size	256	256	256
Pretrain epoch	300	300	400

4.2 Experimental Results

The experiment results are in Table 3. * denotes train, update and merge memory bank simultaneously. R50 denotes use resnet50 backbone. As the Table 3 shows, the training process proposed in this paper can get accuracy improvement on Cifar10 and Imagenet-10 dataset. The more complex backbone also improves the results. The contrastive model with memory bank clustering method have 4% accuracy higher than the mainstream method TSUC [11] on Cifar10 dataset.

Table 3. Comparison results with other sota models on Cifar10, Stl-10, Imagenet-10 dataset, MBC denotes our memory bank clustering model in this paper

Model	Cifar10 (acc-%)	Stl-10 (acc-%)	Imagenet-10 (acc-%)
JULE [28]	27.2	27.7	30.0
DEC [29]	30.1	35.9	38.1
DAC [5]	52.2	47.0	52.7
DCCM [30]	62.3	48.2	71.0
GATCluster [31]	61.0	58.3	76.2
CC [32]	79.0	85.0	89.3
TSUC [11]	81	66.5	N/A
IDFD [19]	81.5	75.6	95.4
SCAN [33]	87.6	76.7	N/A
SPICE [34]	92.6	93.8	96.7
MBC*	80.3	82.1	84.9
MBC	85.69	81.4	90.3
MBC (R50)	91.7	87.9	93.4

Other losses are also tested in the experiment. We modify additive angular margin loss [8] and dsam loss [16] for optimizing our encoder M with memory bank centers C and test on cifar10 dataset. The experiment result show in Table 4, the result shows that our loss can achieve a good accuracy for image classification task. l_b means only use our batch loss to optimize the model without memory bank. The premise of using memory for training is use l_b to pretrain the model, so no result for only use memory bank clustering loss, but we apply our loss directly to memory bank feature without clustering the result shows in MBC-nc.

Table 4. Ablation experiment result

Loss	Cifar10-Acc (%)	Backbone
Dsam [16]	83.5	resnet 18
Arc face [8]	84.92	resnet 18
MBC-l_b	81.7	resnet 18
MBC-nc	84.52	resnet 18
MBC	85.69	resnet 18

As Fig. 6 shows, we use the t-SNE [2] algorithm to map High-dimensional data points to a 2D plane. With the continuous optimization of the encoder network, We can see the number of centers in the memory bank gradually decrease are getting closer to the number of categories used in training data. The distance between different centers is getting farther with model optimization. The number of red points in Fig. 6 denotes the hypothetical class center. In the training epoch, the center in the memory bank can be appropriately and reasonably merged as a new class center of similar features or training data. The distance between two adjacent red dots gradually increases, proving that our network learns the difference between different classes well.

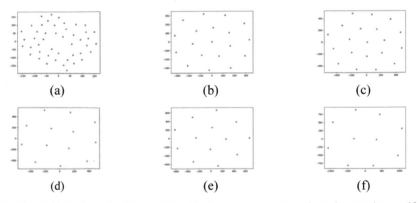

(a)   (b)   (c)

(d)   (e)   (f)

Fig. 6. The t-SNE [2] result, this t-SNE method can map memory clustering results on cifar10 dataset to 2D plane distribution. This figure shows the illustration of the memory centers distribution during 200, 500, 800, 1000, 1200 and 1400 epoch respectively

5 Conclusion

This paper proposes a new network architecture for self-supervised contrastive Learning based on a dynamically updated and memory bank with feature clustering. A feature clustering algorithm is designed for the memory bank. This paper also designed a contrastive method to construct sample pairs between batch features and memory centers and applied contrastive loss with memory bank centers to optimize the encoder network. As the experiments result on the Cifar10, Stl10 and Imagenet-10 dataset shows, the Memory Bank Clustering method can provide positive and negative samples for the contrastive network, and the model can learn a good feature representation.

References

1. Bachman, P., Hjelm, R.D., Buchwalter, W.: Learning representations by maximizing mutual information across views. arXiv preprint arXiv:1906.00910 (2019)
2. Belkina, A.C., Ciccolella, C.O., Anno, R., Halpert, R., Spidlen, J., Snyder-Cappione, J.E.: Automated optimized parameters for T-distributed stochastic neighbor embedding improve visualization and analysis of large datasets. J. Nat. Commun. **10**, 1–12 (2019)
3. Bellet, A., Habrard, A., Sebban, M.: A survey on metric learning for feature vectors and structured data. arXiv preprint arXiv:1306.6709 (2013)
4. Caron, M., Bojanowski, P., Joulin, A., Douze, M.: Deep clustering for unsupervised learning of visual features. In: Ferrari, V., Hebert, M., Sminchisescu, C., Weiss, Y. (eds.) Computer Vision – ECCV 2018. LNCS, vol. 11218, pp. 139–156. Springer, Cham (2018). https://doi.org/10.1007/978-3-030-01264-9_9
5. Chang, J., Wang, L., Meng, G., Xiang, S., Pan, C.: Deep adaptive image clustering. In: Proceedings of the IEEE International Conference on Computer Vision, pp. 5879–5887 (2017)
6. Chen, T., Kornblith, S., Norouzi, M., Hinton, G.: A simple framework for contrastive learning of visual representations. In: International Conference on Machine Learning, pp. 1597–1607 (2020)
7. Chen, X., Fan, H., Girshick, R., He, K.: Improved baselines with momentum contrastive learning. arXiv preprint arXiv:2003.04297 (2020)
8. Deng, J., Guo, J., Xue, N., Zafeiriou, S.: ArcFace: additive angular margin loss for deep face recognition. In: Proceedings of the IEEE/CVF Conference on Computer Vision and Pattern Recognition, pp. 4690–4699 (2019)
9. Pytorch lightning. https://github.com/PyTorchLightning/pytorch-lightning
10. Hadsell, R., Chopra, S., Lecun, Y.: Dimensionality reduction by learning an invariant mapping. In: 2006 IEEE Computer Society Conference on Computer Vision and Pattern Recognition (CVPR 2006), pp. 1735–1742. IEEE (2006)
11. Han, S., Park, S., Park, S., Kim, S., Cha, M.: Mitigating embedding and class assignment mismatch in unsupervised image classification. In: Vedaldi, A., Bischof, H., Brox, T., Frahm, J.M. (eds.) Computer Vision – ECCV 2020. LNCS, vol 12369. Springer, Cham (2020). https://doi.org/10.1007/978-3-030-58586-0_45
12. He, K., Fan, H., Wu, Y., Xie, S., Girshick, R.: Momentum contrast for unsupervised visual representation learning. In: Proceedings of the IEEE Computer Society Conference on Computer Vision and Pattern Recognition, pp. 9729–9738 (2020)
13. He, K., Zhang, X., Ren, S., Sun, J.: Deep residual learning for image recognition. In: Proceedings of the IEEE Conference on Computer Vision and Pattern Recognition, pp. 770–778 (2016)

14. Jaiswal, A., Babu, A.R., Zadeh, M.Z., Banerjee, D., Makedon, F.: A survey on contrastive self-supervised deep metric learning: a survey learning. J. Tech. **9**, 2 (2021)
15. Ji, X., Henriques, J.F., Vedaldi, A.: Invariant information clustering for unsupervised image classification and segmentation. In: Proceedings of the IEEE/CVF International Conference on Computer Vision, pp. 9865–9874 (2019)
16. Kong, J., Cheng, Y., Zhou, B., Li, K., Xing, J.: DSAM: a distance shrinking with angular marginalizing loss for high performance vehicle re-identification. arXiv preprint arXiv:2011.06228 (2020)
17. Krizhevsky, A., Hinton, G.: Learning multiple layers of features from tiny images (2009)
18. Misra, I., Maaten, L.V.D.: Self-supervised learning of pretext-invariant representations. In: Proceedings of the IEEE/CVF Conference on Computer Vision and Pattern Recognition, pp. 6707–6717 (2020)
19. Tao, Y., Takagi, K., Nakata, K., Center, C.R.: Clustering-friendly representation learning via instance discrimination and feature decorrelation. arXiv preprint arXiv:2106.00131 (2021)
20. Wang, X., Zhang, H., Huang, W., Scott, M.R.: Cross-batch memory for embedding learning. In: Proceedings of the IEEE/CVF Conference on Computer Vision and Pattern Recognition, pp. 6388–6397 (2020)
21. Weston, J., Chopra, S., Bordes, A.: Memory networks. arXiv preprint arXiv:1410.3916 (2014)
22. Wu, Z., Xiong, Y., Yu, S.X., Lin, D.: Unsupervised feature learning via nonparametric instance discrimination. In: Proceedings of the IEEE Conference on Computer Vision and Pattern Recognition, pp. 3733–3742 (2018)
23. Ye, M., Zhang, X., Yuen, P.C., Chang, S.F.: Unsupervised embedding learning via invariant and spreading instance feature. In: Proceedings of the IEEE/CVF Conference on Computer Vision and Pattern Recognition, pp. 6210–6219 (2019)
24. Zhang, R., Isola, P., Efros, A.A.: Colorful image colorization. In: Leibe, B., Matas, J., Sebe, N., Welling, M. (eds.) ECCV 2016. LNCS, vol. 9907, pp. 649–666. Springer, Cham (2016). https://doi.org/10.1007/978-3-319-46487-9_40
25. Deng, J., Dong, W., Socher, R., Li, L.J., Li, K., Fei-Fei, L.: ImageNet: a large-scale hierarchical image database. In: CVPR (2009)
26. Coates, A., Ng, A., Lee, H.: An analysis of single-layer networks in unsupervised feature learning. In: Proceedings of the Fourteenth International Conference on Artificial Intelligence and Statistics, pp. 215–223. JMLR Workshop and Conference Proceedings (2011)
27. Yadan, O.: Hydra - a framework for elegantly configuring complex applications. J. Github **2**, 5 (2019)
28. Yang, J.W., Parikh, D., Batra, D.: Joint unsupervised learning of deep representations and image clusters. In: CVPR, pp. 5147–5156 (2016)
29. Xie, J., Girshick, R., Farhadi, R.: Unsupervised deep embedding for clustering analysis. In: ICML 2016, pp. 478–487. JMLR.org (2016)
30. Wu, J., et al.: Deep comprehensive correlation mining for image clustering. In: ICCV, pp. 8150–8159 (2019)
31. Niu, C., Zhang, J., Wang, G., Liang, J.: GATCluster: self-supervised Gaussian-attention network for image clustering. In: Vedaldi, A., Bischof, H., Brox, T., Frahm, J.-M. (eds.) ECCV 2020. LNCS, vol. 12370, pp. 735–751. Springer, Cham (2020). https://doi.org/10.1007/978-3-030-58595-2_44
32. Li, Y., Hu, P., Liu, Z., Peng, D., Zhou, J.T., Peng, X.: Contrastive clustering. In: AAAI (2021)
33. Van Gansbeke, W., Vandenhende, S., Georgoulis, S., Proesmans, M., Van Gool, L.: Scan: learning to classify images without labels. In: Vedaldi, A., Bischof, H., Brox, T., Frahm, J.-M. (eds.) ECCV 2020. LNCS, vol. 12355, pp. 268–285. Springer, Cham (2020). https://doi.org/10.1007/978-3-030-58607-2_16
34. Niu, C., Wang, G.: Spice: semantic pseudo-labeling for image clustering. arXiv preprint arXiv:2103.09382 (2021)

Robust Visual Question Answering Based on Counterfactual Samples and Relationship Perception

Hong Qin[1,2], Gaoyun An[1,2(✉)], and Qiuqi Ruan[1,2]

[1] Institute of Information Science, Beijing Jiaotong University, Beijing 100044, China
{20120310,gyan,qqruan}@bjtu.edu.cn
[2] Beijing Key Laboratory of Advanced Information Science and Network Technology, Beijing 100044, China

Abstract. Traditional visual question answering algorithms based on relationship perception help answer questions by modeling the relationship in the input image. Although better visual question answering performance can be obtained, the model learns the language deviation of the image appearance during training and performs slightly worse on test sets with different data distributions. A model based on Counterfactual Samples and Relationship Perception (CSRP) is proposed by us to solve this problem. The counterfactual sample generation mechanism can generate a large number of counterfactual samples by shielding key objects, forcing the model to focus on key objects to answer questions. Counterfactual samples as feature enhancements can reduce the learning appearance language bias during training. And the relationship between image objects perceives semantics. Extensive experiments on the VQA-CP v2 and VQA V2 datasets demonstrate that our proposed model outperforms most state-of-the-art methods.

Keywords: Relationship perception · Counterfactual samples · Visual question answering

1 Introduction

In recent years, deep learning is booming. At the same time, natural language processing and computer vision have also made qualitative leaps. Visual question answering (VQA) task that combines image and text has also attracted much attention. Many large datasets such as VQA V2 [11] and VQA-CP v2 [2] have also been released to promote the proposal of numerous VQA algorithms. For the VQA task, under the multimodal input of a given image and image-based questions, it aims to obtain answers to the questions according to the content of the image.

In computer vision tasks, relationships between objects in images are extremely important for modeling algorithms. Early work [4, 8, 10] also made some explorations on

Supported by the Fundamental Research Funds for the Central Universities 2021YJS043, and the National Natural Science Foundation of China 62072028 and 61772067.

this. Divvala et al. [8] applied the relationship information between the objects contained in the image to the task of object detection. The modeling of co-occurrence and relative position information between image objects [10] has also been successfully applied to object classification tasks. Different sizes of relationships between image objects [4] can also be modeled as advanced visual features. The semantic and spatial relationship between images [26] is used to model image description tasks to obtain sentences similar to human language descriptions. Similarly, the relationships between these image objects have also been well applied to the VQA task. Li et al. [14] proposed a relationship perception model ReGAT, which defined the spatial and semantic relationships between objects as explicit relationships, and defined the high-level relationships captured by the fully connected graphs between modeled objects as implicit relationships. Li et al. [14] used graph attention network to model the explicit relationships and implicit relationships between image objects, and achieved excellent VQA performance.

Although many algorithms [1, 9, 14, 15, 17] have been proposed in VQA tasks, due to the large amounts of deviations in their datasets, the above algorithms often learn the pseudo-correlation of the image surface and cannot learn the real relationships between the visual characteristics. For instance, for the question "What is the color of the banana in this picture?", the model directly answers "yellow" according to the learned deviation, but the banana in the picture may also be black or cyan. These learned language biases will perform poorly on test sets with different question answering distributions. The current methods to reduce language bias mainly include: (1) Obtain a more balanced dataset. For example, the proposal of the VQA-CP v2 dataset divides the training set and the test set into completely different distributions, which greatly reduces the data deviation and alleviates the long-tail distribution problem of different types of data. (2) Design a more suitable network structure. Chen et al. [6] proposed a counterfactual sample generation model CSS, which generates a good deal of counterfactual samples by masking the key objects of images or the keywords of questions, so that the model is forced to focus on the key objects to answer the question and generate better vision question answering results.

A visual question answering model based on Counterfactual Samples and Relational Perception (CSRP) was raised by us to solve the above problems. Our model integrates the counterfactual sample generation mechanism with the relational perception module and has a new loss function. The counterfactual sample generation mechanism can cover up the key objects in the image or cover up the keywords in the question or replace them with synonyms to generate a great many counterfactual samples to enhance image features, which reduces the long tail distribution problem in the dataset and forces the model to focus on the key object. The relationship perception mechanism can learn more fine-grained visual features by using explicit relationships such as spatial location information between image objects, semantic relationships, and perceiving dynamic implicit relationships between image regions. These visual features include deeper relationships between objects to answer complex semantic questions. The fusion of counterfactual samples and relationship perception can perceive a variety of complex relationships between image objects to enhance visual features while using counterfactual sample generation mechanisms to balance datasets, reduce language bias, and obtain more accurate results for complex semantic problems result. Experimental results demonstrate that our model can surpass most current models on the VQA-CP v2 dataset and VQA V2 dataset.

In conclusion, we have made the following contributions:

- We put forward a VQA model based on Counterfactual Samples and Relationship Perception (CSRP). And we design a new loss function for our model. This is the first time that counterfactual samples are combined with relational perception. The graph convolutional attention network is used to perceive the complex semantic relationship between image objects. The counterfactual sample generation mechanism generates numerous counterfactual samples to reduce the learned language bias.
- A good deal of experimental results shows that this method surpasses most of the advanced models on the VQA-CP v2 dataset and has made significant progress. Our CSRP model reduces the deviation of the experimental results between the VQA-CP v2 dataset and the VQA V2 dataset.

2 Related Work

2.1 Visual Question Answering

The general idea of the current visual question answering task is to use the image feature extractor (usually CNN) and the question feature extractor (usually RNN) to encode the input images and the questions respectively, and then to fuse the multimodal features of the images and the questions. Finally, the fusion feature is used as the input of the classifier to obtain the VQA result. Some works [16, 20, 27] explored the use of an attention mechanism as an image feature extractor to obtain problem-related image regions. The new problem feature extractor [15, 16] was also designed to explore the visual features related to the problem. We directly encode the relationship between image objects as fine-grained visual features to be applied to the visual question answering task.

2.2 Counterfactual Sample Synthesis Mechanism for VQA

Some recent research works [1, 6, 18] began to focus on the direction of counterfactual sample generation to enhance the input image and problem features to reduce language bias. Agarwal et al. [1] proposed the use of a GAN-based [30] model to re-synthesize images from a causal point of view. Inspired by [6], the counterfactual sample synthesis mechanism we adopted uses the visual features extracted by Faster R-CNN to mask several key objects to obtain the counterfactual image or mask or replace the problem features encoded by the problem to obtain the counterfactual problem. This method only cuts and replaces the original image or problem, which is more convenient and simpler.

2.3 Relationship Perception in VQA

We define the spatial position relationship, size relationship, co-occurrence relationship, and semantic relationship representing action interaction between image objects as explicit relationships. At the same time, there are some high-level implicit relationships between image regions that are not visible but can be learned through the network.

Generally [13, 22] dynamically captured implicit relationships by constructing a fully connected graph of input features. Santoro et al. [22] used the attention mechanism to model pairwise relationships end-to-end. Our relationship perception includes explicit relationships and implicit relationships. By using graph attention networks to model explicit relationships, different nodes can be assigned different weights. Implicit relationships are realized by constructing a fully connected relationship graph. We model explicit and implicit relationships to obtain high-level features of the image, which helps the VQA task.

3   Method

We first make a simple definition of the VQA task: based on taking VQA as a multi-classification task, given the input image I, the question Q and the standard answer a $\in A$, our goal is to pass the mapping function $f_{vqa}: I \times Q \rightarrow [0, 1]^{|A|}$. The predicted answer a* obtained by learning can approximate the real answer a to the greatest extent.

$$a* = f_{vqa}(V, Q) = P_{vqa}(a|I, Q). \tag{1}$$

The basic framework of our proposed CSRP model is revealed in Fig. 1. We propose a relationship perception module and a counterfactual sample generation module based on the UpDn model [3]. The input image is passed through the Faster R-CNN encoder to extract visual features, and the input question is passed through the problem encoder to extract semantic features. Then the extracted visual and semantic features are fused, and the answer to the question is obtained through the classifier.

Our CSRP model first adds a relational perception module after the image encoder and the question encoder and combines the visual features and semantic features into visual semantic pairs to construct the relation graph, so that the relation graph will be based on the question semantics. Relations related to the problem are assigned greater weight. The gained relationship graph features and the text features acquired by the question encoder are further multimodally fused in the fusion module of UpDn. After passing through the UpDn classifier module, the predicted answer obtained in the above process is combined with the input image and the question feature as the input of our counterfactual sample generation mechanism. The counterfactual sample generation mechanism will acquire counterfactual samples by overspreading key objects of the original image or covering keywords in the question. The generated counterfactual samples and original samples will be used as input samples to the VQA model to obtain our final predicted answer.

3.1   Relationship Perception Network

The content of the blue box in Fig. 1 represents our relationship perception module. All object features and problem features of the input image will be combined into pairs to construct an explicit relationship graph and an implicit relationship graph to obtain high-level visual features. Next, we will introduce three parts: implicit graph construction, explicit graph construction, and question-adaptive relational encoder.

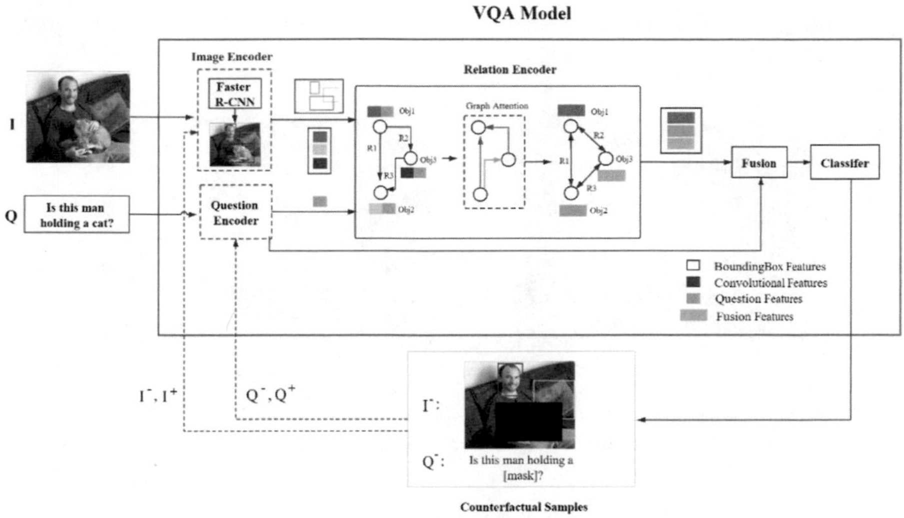

Fig. 1. A model framework based on Counterfactual Samples and Relationship Perception (CSRP). Among them, I, I^+, I^- represent the original image, factual image and counterfactual image, Q, Q^+, Q^- represent the original question, factual question and counterfactual question. O_{bj} represents the object, R represents the relationship.

Implicit Graph Construction. The implicit relationship refers to the high-level dynamic relationship existing between the object regions of the image. Although it is not visible, it can be obtained through learning. Implicit relations are extremely efficient for answering complex questions in VQA. Constructing implicit relationships into an implicit graph can facilitate updating with the help of graph structure. For each object v_i obtained by Faster R-CNN encoding of the image, $i = 1, 2, \ldots k$, we regard each object v_i as a vertex, which can form a fully connected undirected graph $G = (V, E)$, where V is the set of all object vertices, and E is the set of $k \times (k - 1)$ edges. This fully connected undirected graph can be expressed as an implicit graph, and each edge in E is used to learn the implicit relationship between different objects.

Explicit Graph Construction. Pre-trained classifiers are used to extract different relationships between input objects and obtain prior knowledge about the edges between objects. For the implicit relational graph, we can cut out the edges without relevant knowledge to convert the implicit graph to the explicit graph based on prior knowledge. For the explicit relationship between image objects, we explore the spatial relationship and semantic relationship. Inspired by [14], we model spatial relations and semantic relations as spatial graphs and semantic graphs. The spatial graphs can be expressed as $<object_i - predicate - object_j>$, where *predicate* represents the spatial position relationship of *object_i* relative to *object_j*. As shown in the left picture of Fig. 2, $<horse - over - chair>$ means the toy horse is over the chair. The spatial diagram relationship must be symmetrical. The semantic graphs can be expressed as $<subject_i - predicate - object_j>$. As shown in the right picture of Fig. 2, $<man - surfing - surfboard>$

means that the man is surfing on a surfboard. The relationship between semantic graphs is unidirectional.

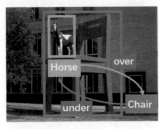
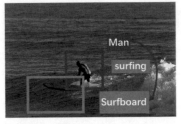

(a): Spatial Relation   (b): Semantic Relation

Fig. 2. Examples of spatial relations and semantic relations. The arrow indicates the direction of the relationship, and the text in the middle of the arrow indicates the specific relationship.

Question-Adaptive Relational Encoder. After constructing the visual and textual features into explicit and implicit graphs, the problem-adaptive relational encoder is used to extract the dynamic relationship between these objects. The details are as follows: we combine question features with visual relation features to design a question-based relational encoder and input the semantic information carried in the question into the relational encoder. A relational graph can be learned in line with the importance of the question, where the edges related to the question will have greater significance. This is done by embedding the problem feature q with the visual feature v_i, $i = 1, 2,... k$. As shown in formula 2:

$$v'_i = [v_i||q] \quad i = 1, \cdots, k \tag{2}$$

After feature embedding, the self-attention mechanism is used for each object in the relationship graph to learn the hidden fine-grained relationship features $\{v^*_i\}^k_{i=1}$. Among them:

$$v^*_i = \sigma\left(\sum_{j \in N_i} a_{ij} \cdot W v'_j\right). \tag{3}$$

In formula 3, σ represents a nonlinear activation function. a_{ij} is the attention weight coefficient of the relation graph, and different relation graphs have different coefficients. W is the projection matrix, $W \in R^{d_h \times (d_q + d_v)}$. We continue to fuse the high-level relationship feature v^*_i obtained by the above operation with the original question feature q in the feature fusion layer, and then pass it to the classifier to obtain the predicted answer result.

3.2 Counterfactual Sample Generation Mechanism

Our counterfactual sample module is shown in the green box in Fig. 1. After the relationship perception based UpDn algorithm, we can get the triple (I, Q, a^*), where a^* is the predicted answer we get. Inputting the above triples into the counterfactual sample generation mechanism to obtain image triples (I, I^+, I^-) and question triples (Q, Q^+, Q^-), and use the obtained image and question features as the input of the VQA model. According to the more evenly distributed input data, we can get a better visual question answering performance. The process is shown in Algorithm 1. First, use the original sample to train the model to get the triples $(I, Q, a*)$, and execute the image counterfactual or problem counterfactual according to the generated random numbers, use the generated counterfactual image or question together with the original sample as the input of the training model to continue training. Next, we will give a detailed introduction to the image counterfactual generation mechanism and the problem counterfactual generation mechanism. We also introduce the loss function. In the following, for convenience, we use a to denote $a*$.

Algorithm 1   CSRP Model Based on VQA Task

1: **function** CSRP (I, Q, a)
2:       def function VQA(I, Q a)
3:            $V \leftarrow e_v(I)$
4:            $Q \leftarrow e_q(Q)$
5:            $P_{vqa} \leftarrow f_{vqa}(V, Q)$
6:            $Loss \leftarrow XE(P_{vqa}, a)$
7:       end function
8:       $V, Q, P_{vqa}(a) \leftarrow VQA(I, Q, a)$
9:       **if** $cond \geqslant \xi$ **then**
10:           execute *Image-Counterfactual*
11:     **else**
13:           execute *Problem-Counterfactual*
14:       $VQA(I, Q^-, a^-)$ **Or** $VQA(I^-, Q, a^-)$
15: **end function**

Image Counterfactual Generation Mechanism. Based on the visual objects in the text question and answer pair (Q, a), not all visual objects I are related to the question or answer. Therefore, we crop all visual objects according to the questions and answers to delete irrelevant visual objects and reduce the scope of related visual objects. In a specific implementation, we first perform part-of-speech tagging on (Q, a) to extract the noun entity, calculate the cosine similarity between the noun and the object category, and use the first N image objects with the highest cosine similarity score as our benchmark object set I. Then calculate the importance of each entity object to the answer on the benchmark object set.

We use a modified Grad-Cam [22] to calculate the relative importance of a single object to the ground truth. The calculation formula for the relative importance of the i-th object is:

$$s(a, v_i) = S\left(P_{vqa}(a), v_i\right) = \left(\nabla v_i P_{vqa}(a)\right)^T 1. \tag{4}$$

where v_i represents the i-th object feature, and $P_{vqa}(a)$ represents the predicted probability of answer a obtained through the VQA model. 1 is a vector all ones. After getting $s(a, v_i)$, we sort the object set in descending order according to it, and take the first K objects in the object set as the key object set I^+. The selection of K value is dynamic. I^- is generated by subtracting I^+ from the set of objects I. Figure 3 displays a case of how we generate I^+ and I^-. For the counterfactual image, it forms a triple (I^-, Q, a^-) with the original question and the counterfactual answer. The a^- is generated by adding negative words to the original answer, such as "Not".

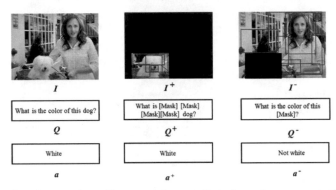

Fig. 3. Examples of counterfactual images and counterfactual questions generated by the original image question pair.

Problem Counterfactual Generation Mechanism. For the problem counterfactual, its generation steps are similar to the image counterfactual generation mechanism. First, compute the significance of the i-th word feature in question to ground truth answer:

$$s(a, w_i) = S(P_{vqa}(a), w_i) = (\nabla w_i P_{vqa}(a))^T 1. \tag{5}$$

Then divide the question into words that mark the question type and the remaining words. As shown the question "What is the color of this dog?" in Fig. 3, the word marked with the question type is: "What is". We select the top M words with the highest $s(a, w_i)$ among the remaining words as key words, and replace the key words we get with special characters "[Mask]" to get the counterfactual question sentence Q^-. As in the example, get Q^-: "What is the color of this [Mask]?". Then Q^+ is to remain the key words we achieved in Q^- in the remaining words under the condition that the problem type remains unchanged, and replace all the remaining words with "[Mask]".

3.3 Loss Functions

Generally, during VQA training, this task is regarded as a multi-classification task. And employ the soft maximum cross-entropy function for training:

$$Lvqa = \frac{-1}{N} \sum_{i=1}^{N} \log(softmax(f_{vqa}(X_i), a_i)). \tag{6}$$

And our training model has made some improvements on this basis. Our $f_{vqa}(X_i)$ can be computed by:

$$f_{vqa}(X_i) = \alpha \, P_{sem}(X = X_i) + \beta \, P_{spa}(X = X_i) + (1 - \alpha - \beta) \, P_{imp}(X = X_i). \tag{7}$$

where $P_{sem}(X = X_i)$, $P_{spa}(X = X_i)$, and $P_{imp}(X = X_i)$ represent the predicted probability of the answer gained using the pre-trained semantic relationship model, spatial relationship model, and implicit relationship model. α, β are training hyperparameters.

During training, the traditional cross-entropy loss function cannot improve the data with the long-tailed distribution. Inspired by the good performance of focal loss proposed in [28] on object detection tasks, we use this loss function in VQA tasks. The focal loss is based on the standard cross-entropy loss function. Aiming at the problem of unbalanced data category difficulty, the design function pays more attention to difficult-to-classify samples. As shown in the following formula:

$$Lvqa = \frac{-1}{N} \sum_{i=1}^{N} \{(1 - softmax(f_{vqa}(X_i),)^{\gamma} \log(softmax(f_{vqa}(X_i), a_i))\}. \quad (8)$$

where γ is called the focusing parameter, $\gamma \geq 0$. The purpose of adding this modulation factor is to assign greater weight to difficult-to-classify samples to balance the difficulty of the samples.

4 Experiment

4.1 Dataset

Our model has conducted extensive experiments on two challenging benchmark datasets, the VQA V2 dataset, and the VQA-CP v2 dataset. The VQA V2 dataset contains real images from the MSCOCO dataset. It consists of 82783 training image samples, 40504 validation set sample images and 81434 test sample images, and includes multiple open questions and answers provided by real people. The VQA-CP v2 dataset rest on the VQA V2 dataset. The training set and the validation set are re-divided to punish the type of answer given based on the deviation of the question. The VQA-CP v2 dataset is proposed to simulate the distribution of the actual application dataset.

4.2 Parameter Setting

In our model implementation, we use predefined classifiers to extract prior knowledge about explicit relationships. Specifically, Resnet-101 is used to extract the bounding box features and target detection features of the original input image. For the counterfactual image module, we select 9 basic objects with the highest cosine similarity score as our basic object set. In the counterfactual question module, only using "[Mask]" to replace one key object works best. Our model is implemented in Pytorch, and the batch size is set to 256. Our loss function is focal loss. We use the Adamax optimizer for training. And in terms of learning rate adjustment, we apply a warm-up strategy. In order to reduce the parameters, dropout operation and normalization are performed after each linear layer.

4.3 Performance on the VQA-CP v2 Dataset

We verify the experimental performance of our proposed CSRP model on the VQA CP v2 dataset and VQA V2 dataset and compare them with the advanced models of the two datasets. On the VQA-CP v2 dataset, the models we compared are divided into four categories: The first category is SAN-based methods [25], for example, GVQA [2]. The second basic model is a non-reorganized method, including CF [23] and CF+GS [23]. The third method is based on UpDn [3], which is currently the more mainstream method, including RUBi [5], LMH [7], etc. The fourth method is based on the VQA reasoning model, including multi-modal reasoning MuRel [5] and causal-based reasoning CIKD [19]. Our CSRP model can achieve 58.55% of the experimental performance, which is better than the other models in Table 1. And compared with 57.59% of the second-ranked SSL model in the table, our model improves the experimental performance by 0.96%. On the Num problem, we obtained the best result of 50.80%. The experimental effect of the CSRP model we proposed has reached the first-class level in the VQA-CP v2 dataset.

4.4   Performance on the VQA V2 Dataset

Table 2 demonstrates the experimental results of our CSPR model on the VQA V2 dataset. On the VQA V2 dataset, we compare seven advanced models including GVQA and RUBi. The experimental results indicate that our model can get relatively leading experimental results on this dataset, and the accuracy rate is as high as 62.37%. Although it is 1.11% lower than the experimental result of UpDn, our experimental result exceeds the other models in Table 2. In general, our CSRP model can still achieve high results on the VQA V2 dataset.

4.5   Ablation Experiment of CSRP Model

To explore the significance of our proposed model, we conducted ablation experiments to compare our CSRP model with the CSS model using only the counterfactual sample synthesis mechanism and the ReGAT model using only the relational perception module. Table 3 displays the results of our experiment. On the VQA-CP v2 dataset, CSRP achieves a good performance of 58.55%. Compared with the ReGAT model, the experimental results have increased from 39.60% to 58.55%, an increase of 18.95%. And increased the CSS model from 57.74% to 58.55%, an increase of 0.81%. Such excellent results can be achieved because the CSRP model can combine the advantages of the counterfactual sample synthesis mechanism and the relational perception module. On the VQA V2 dataset, we get a result of 62.37%. Compared with the CSS model, our experimental results increased by 2.46%, which indicates that the relational perception learned by our CSRP module can further improve the performance of VQA. Compared with ReGAT, our model reduces by 1.21%. This may be because the counterfactual sample generation mechanism interferes with the surface deviation that the model learns on the dataset, thereby reducing performance. And the experimental results in Table 3 display that the CSRP model can reduce the result deviation between these two datasets to 3.82%. This is compared with the 23.98% data deviation of the ReGAT model, indicating that our CSRP model can effectively learn the true correlation of the dataset to different distributed datasets. And effectively reduce the language deviation in the VQA dataset, and enhance the migration ability of the model.

Table 1. Experimental results on the VQA-CP v2 dataset, * indicates that we have re-implemented the experimental results of the model, and the best experimental results are shown in bold.

VQA-CP v2 *test*							
Model	Venue	Expl	Overrall	Yes/No	Num	Other
SAN [25]	CVPR'16		24.96	38.35	11.14	21.74
GVQA [2]	CVPR'18		31.30	57.99	13.68	22.14
Unshuffing [24]	ECCV'20		42.39	47.72	14.43	47.24
+CF [23]	ECCV'20	HAT	46.00	61.30	15.60	46.00
+CF+GS [23]	ECCV'20	HAT	46.80	64.50	15.30	45.90
UpDn [3]	CVPR'18		39.74	42.27	11.93	46.05
+ReGAT*[14]	ICCV'19		39.60	42.57	12.08	45.60
+AREG [21]	NeurIPS'18		41.17	65.49	15.48	35.48
+GRL [12]	ACL'19		42.33	59.74	14.78	40.76
+RUBi [5]	NeurIPS'19		44.23	67.05	17.48	39.61
+LMH [7]	EMNLP'19		52.01	72.58	31.12	46.97
+SSL [29]	AAAI'20		57.59	86.53	29.87	**50.03**
MuRel [5]	CVPR'19		39.54	42.85	13.17	45.04
CIKD [19]	MMAsia'21		54.05	**90.01**	15.10	45.88
CSRP (Ours)			**58.55**	88.22	**50.80**	45.13

Table 2. The experimental results on the VQA V2 dataset, the results of best experiment are shown in bold.

VQA V2 *val*				
Model	Overrall	Yes/No	Num	Other
GVQA [2]	48.24	72.03	31.17	34.65
RUBi [5]	50.56	49.45	41.02	53.95
CSS [6]	59.91	73.25	39.77	55.11
Unshuffling [24]	61.08	78.32	**42.16**	52.81
LMH [7]	61.64	77.85	40.03	55.04
CIKD [19]	61.29	76.34	40.20	55.43
UpDn [3]	**63.48**	**81.18**	42.14	**55.56**
CSRP (Ours)	62.37	80.72	41.58	53.93

Table 3. The ablation experiment of the CSRP model, * indicates that we have re-implemented the experimental results of the model. The best experiment is highlighted in bold.

Model	VQA V2 *val*				VQA-CP v2 *test*				GAP
	Overrall	Y/N	Num	Other	Overrall	Y/N	Num	Other	
ReGAT*	**63.58**	**81.94**	43.97	54.86	39.60	42.57	12.08	45.60	23.98
CSS*	59.91	73.25	39.77	**55.11**	57.74	83.18	47.59	**47.19**	2.17
CSRP kik (Ours)	62.37	80.72	**41.58**	53.93	**58.55**	**88.22**	**50.80**	45.13	3.82

5   Conclusion

To improve the performance of visual question answering and reduce the language bias learned in the training dataset, we put forward a model based on Counterfactual Samples and Relational Perception (CSRP). This is the first time that the counterfactual sample synthesis mechanism has been integrated with the relational perception module. The model can learn more visually fine-grained relational features through relational perception. Moreover, it can obtain counterfactual samples by shielding the key objects of the original sample, forcing the model to focus on answering questions most relevant to the object. This can enhance the characteristics of the input data and reduce the deviation of the dataset. Our proposed model can surpass most other models on the VQA-CP v2 data set and VQA V2 data set, which proves the effectiveness of our model.

References

1. Agarwal, V., Shetty, R., Fritz, M.: Towards causal VQA: revealing and reducing spurious correlations by invariant and covariant semantic editing. In: Proceedings of the IEEE/CVF Conference on Computer Vision and Pattern Recognition, pp. 9690–9698 (2020)
2. Agrawal, A., Batra, D., Parikh, D., et al.: Don't just assume; look and answer: overcoming priors for visual question answering. In: Proceedings of the IEEE Conference on Computer Vision and Pattern Recognition, pp. 4971–4980 (2018)
3. Anderson, P., He, X., Buehler, C., et al.: Bottom-up and top-down attention for image captioning and visual question answering. In: Proceedings of the IEEE Conference on Computer Vision and Pattern Recognition, pp. 6077–6086 (2018)
4. Biederman, I., Mezzanotte, R.J., Rabinowitz, J.C.: Scene perception: detecting and judging objects undergoing relational violations. Cogn. Psychol. **14**(2), 143–177 (1982)
5. Cadene, R., Ben-Younes, H., Cord, M., et al.: Murel: multimodal relational reasoning for visual question answering. In: Proceedings of the IEEE/CVF Conference on Computer Vision and Pattern Recognition, pp. 1989–1998 (2019)
6. Chen, L., Yan, X., Xiao, J., et al.: Counterfactual samples synthesizing for robust visual question answering. In: Proceedings of the IEEE/CVF Conference on Computer Vision and Pattern Recognition, pp. 10800–10809 (2020)
7. Clark, C., Yatskar, M., Zettlemoyer, L.: Don't take the easy way out: ensemble based methods for avoiding known dataset biases. arXiv preprint arXiv:1909.03683 (2019)
8. Divvala, S.K., Hoiem, D., Hays, J.H., et al.: An empirical study of context in object detection. In: 2009 IEEE Conference on Computer Vision and Pattern Recognition, pp. 1271–1278. IEEE (2009)

9. Fan, H., Zhou, J.: Stacked latent attention for multimodal reasoning. In: Proceedings of the IEEE Conference on Computer Vision and Pattern Recognition, pp. 1072–1080 (2018)
10. Galleguillos, C., Rabinovich, A., Belongie, S.: Object categorization using co-occurrence, location and appearance. In: 2008 IEEE Conference on Computer Vision and Pattern Recognition, pp. 1–8. IEEE (2008)
11. Goyal, Y., Khot, T., Summers-Stay, D., et al.: Making the V in VQA matter: elevating the role of image understanding in visual question answering. In: Proceedings of the IEEE Conference on Computer Vision and Pattern Recognition, pp. 6904–6913 (2017)
12. Grand, G., Belinkov, Y.: Adversarial regularization for visual question answering: strengths, shortcomings, and side effects. arXiv preprint arXiv:1906.08430 (2019)
13. Hu, H., Gu, J., Zhang, Z., et al.: Relation networks for object detection. In: Proceedings of the IEEE Conference on Computer Vision and Pattern Recognition, pp. 3588–3597 (2018)
14. Li, L., Gan, Z., Cheng, Y., et al.: Relation-aware graph attention network for visual question answering. In: Proceedings of the IEEE/CVF International Conference on Computer Vision, pp. 10313–10322 (2019)
15. Lu, J., Yang, J., Batra, D., et al.: Hierarchical question-image co-attention for visual question answering. Adv. Neural. Inf. Process. Syst. **29**, 289–297 (2016)
16. Malinowski, M., Doersch, C., Santoro, A., Battaglia, P.: Learning visual question answering by bootstrapping hard attention. In: Ferrari, V., Hebert, M., Sminchisescu, C., Weiss, Y. (eds.) ECCV 2018. LNCS, vol. 11210, pp. 3–20. Springer, Cham (2018). https://doi.org/10.1007/978-3-030-01231-1_1
17. Nam, H., Ha, J.W., Kim, J.: Dual attention networks for multimodal reasoning and matching. In: Proceedings of the IEEE Conference on Computer Vision and Pattern Recognition, pp. 299–307 (2017)
18. Pan, J., Goyal, Y., Lee, S.: Question-conditioned counterfactual image generation for VQA. arXiv preprint arXiv:1911.06352 (2019)
19. Pan, Y., Li, Z., Zhang, L., et al.: Distilling knowledge in causal inference for unbiased visual question answering. In: Proceedings of the 2nd ACM International Conference on Multimedia in Asia, pp. 1–7 (2021)
20. Patro, B., Namboodiri, V.P.: Differential attention for visual question answering. In: Proceedings of the IEEE Conference on Computer Vision and Pattern Recognition, pp. 7680–7688 (2018)
21. Ramakrishnan, S., Agrawal, A., Lee, S.: Overcoming language priors in visual question answering with adversarial regularization. arXiv preprint arXiv:1810.03649 (2018)
22. Santoro, A., Raposo, D., Barrett, D.G.T., et al.: A simple neural network module for relational reasoning. arXiv preprint arXiv:1706.01427 (2017)
23. Teney, D., Abbasnedjad, E., van den Hengel, A.: Learning what makes a difference from counterfactual examples and gradient supervision. In: Vedaldi, A., Bischof, H., Brox, T., Frahm, J.-M. (eds.) ECCV 2020. LNCS, vol. 12355, pp. 580–599. Springer, Cham (2020). https://doi.org/10.1007/978-3-030-58607-2_34
24. Teney, D., Abbasnejad, E., Hengel, A.: Unshuffling data for improved generalization. arXiv preprint arXiv:2002.11894(2020)
25. Yang, Z., He, X., Gao, J., et al.: Stacked attention networks for image question answering. In: Proceedings of the IEEE Conference on Computer Vision and Pattern Recognition, pp. 21–29 (2016)
26. Yao, T., Pan, Y., Li, Y., Mei, T.: Exploring visual relationship for image captioning. In: Ferrari, V., Hebert, M., Sminchisescu, C., Weiss, Y. (eds.) Computer Vision – ECCV 2018. LNCS, vol. 11218, pp. 711–727. Springer, Cham (2018). https://doi.org/10.1007/978-3-030-01264-9_42

27. Zhu, C., Zhao, Y., Huang, S., et al.: Structured attentions for visual question answering. In: Proceedings of the IEEE International Conference on Computer Vision, pp. 1291–1300 (2017)
28. Lin, T.Y., Goyal, P., Girshick, R., et al.: Focal loss for dense object detection. In: Proceedings of the IEEE International Conference on Computer Vision, pp. 2980–2988 (2017)
29. Zhu, X., Mao, Z., Liu, C., et al.: Overcoming language priors with self-supervised learning for visual question answering. arXiv preprint arXiv:2012.11528 (2020)
30. Goodfellow, I., Pouget-Abadie, J., Mirza, M., et al.: Generative adversarial nets. In: Advances in Neural Information Processing Systems 27 (2014)

Prototype Generation Based Shift Graph Convolutional Network for Semi-supervised Anomaly Detection

Tao Cui[1,2(✉)], Wenyu Song[1,2], Gaoyun An[1,2], and Qiuqi Ruan[1,2]

[1] Institute of Information Science, Beijing Jiaotong University, Beijing 100044, China
{20120297,20120313,gyan,qqruan}@bjtu.edu.cn
[2] Beijing Key Laboratory of Advanced Information Science and Network Technology, Beijing 100044, China

Abstract. Semi-supervised network is an important branch in video anomaly detection. Previous methods committed to modeling the common feature or distribution of normal data. With the introduction of the pose graph, the model can focus on the behavior of the human body. However, graph embedded networks suffer from the heavy computational cost and could not accurately predict the distribution of normal data. To better tackle these issues, a prototype generation-based graph convolutional network is proposed for anomaly detection, which introduce shift operation and prototype generation module to obtain the distribution of normal data while simplifying the model. Extensive experiments is implemented on ShanghaiTech dataset, the result (76.7 AUC) shows that the proposed approach outperforms most of mainstream models.

Keywords: Human pose · Shift operation · Graph convolution · Prototype generation

1 Introduction

Video Anomaly Detection (VAD) aims to identify the abnormal video and locate the position of the abnormal video snippet. As one of the essential supporting technologies for understanding human behavior, VAD has been widely used in real-world scenarios, e.g. video surveillance, customs inspection, and medical treatment. In recent years, methods that used deep learning have achieved remarkable results [1, 3, 13, 14].

Although deep learning methods for VAD have been numerously investigated, there are many problems with these methods. On the one hand, the definition of anomaly is ambiguous. For example, holding a knife can be judged as anomalous if the event happens outside the kitchen. This example indicates that the scene information is indispensable for detection process. Thus noises such as light and background will have a critical effect

Supported by the Fundamental Research Funds for the Central Universities 2021YJS044, and the National Natural Science Foundation of China 62072028 and 61772067.
T. Cui and W. Song—Equal contribution.

on the judgement result. To address this problem, Markovitz et al. [1] proposed to use the result of pose estimation directly, which exceedingly improves the performance of the model when facing various scenes. On the other hand, there is a massive gap in the number of normal and abnormal videos in the dataset. Even the abnormal videos, only a few frames or snippets are identified as anomalous. For the characteristics of the data set, some methods [13, 15, 16] were proposed to create compact, efficient, and robust features. These methods are also called the weak supervision method. Others focus on modeling the commonality of normal video clips. These methods only use normal video clips during the training phase, so we refer to these methods collectively as semi-supervised in the rest of the paper. They either utilize a proxy task [1, 3, 6] or predict the distribution of normal videos [4, 7, 8, 11]. Previous work shows that the latter is more practical and more challenging in real-world scenarios.

However, semi-supervised methods faced a common problem, thus the discriminative of normal/abnormal videos. Owing to anomaly video clips do not participate in the training phase. Recently, Lv et al. [3] used a predictive model and proposed a DPU module to map the normal video into prototypes. Different from the method mentioned above, we focus on the human action itself and propose a novel approach that uses prototypes to save and update the distribution of anomaly action.

As discussed above, we put forward a novel Dynamic Prototype based Graph convolutional network for video anomaly detection, which takes advantage of the human pose to boost the robustness of the network and a dynamic prototype unit to map the pattern of normal data to the prototype pool. As illustrated in Fig. 1, our model takes a series of pose graphs as input. Next, we expand the joint point coordinates into RGB patches and send them to the shift graph convolution module to get the features of the video clips. Then, we implement a prototype generation module to capture the distribution of normal prototypes according to joint patches and time information. Finally, we followed [1] and adopt a deep clustering model to calculate anomaly scores. We perform comprehensive experiments on ShanghaiTech [5], achieving frame-level AUC scores of 76.7%.

In summary, we summarize the contribution into the following three points:

First, a novel shift graph convolution module is designed for reducing the calculation amount.

Secondly, a novel Prototype Generation-based Graph convolutional network is proposed for VAD, which locate the pivotal time clips and human body parts. With a simple module added, the performance improves a lot.

Thirdly, the novel network is verified on ShanghaiTech Dataset, which outperforms the current mainstream semi-supervised method.

2 Related Work

2.1 Video Anomaly Detection

Since VAD is always regarded as an unsupervised issue, current approaches that adopt CNN methods typically utilize another task as a proxy to learn some features and representations. For example, Markovitz et al. [1] used frame reconstruct, and Liu et al. [6] uses the frame predict method. Just like the problems faced by other semi-supervised methods, the powerful representation ability of CNN leads to an 'over-generalizing' dilemma. Inspired by MIL, Morais et al. [13] used a weakly-supervised method. Each video is viewed as a set of video clips that will be assigned a label, which greatly reduce the workload of manual labeling. Recently, Park et al. [4] introduced a memory module into semi-supervised task where each item in the memory bank represents the normal prototypical patterns. With an update scheme, diversified normal patterns are preserved, which improves the model's ability to discriminate. Previous work proved that analyzing human body parts at the instance level is of great significance to the study of human behavior [20]. However, none of the methods mentioned above considers human pose. Markovitz et al. [1] firstly took human pose into consideration and achieve inspire improvement. However, the method has two drawbacks, compute costly and feature represent. In our method, we aim to use the pose to train the model and optimization the feature to get a simple and effective method.

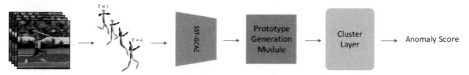

Fig. 1. Overview of Prototype Generation Based Graph Convolutional Network. Firstly, we use pose estimation methods to extract human poses. Then, the shift temporal and spatial graph convolution network (SST-GCN) is implemented to abstract movement features of the human body. Next, we send embedding features into the prototype generation module. After redistributing weights in the time dimension and the key point dimension, a deep clustering layer calculate the final anomaly scores.

2.2 Graph Convolutional Networks

By using the adjacency matrix to assign different weights to the information importance of various neighbors, the graph convolutional network (GCN) can capture more complete feature information. Although it requires more calculations, GCN has a wide range of versatility.

Recently, many approaches were proposed for introducing graph data into computer vision and natural language processing. Following Kipf and Welling [17], Yan et al. [18] and Yu et al. [19] proposed to combine temporal and spatial graph convolution (ST-GCN). To better model the positional relationship of human skeleton nodes, Markovitz et al. [1] proposes to use three parallel graph convolutions to obtain static, global, and attention-based feature information. We follow their spatial attention graph convolution module, illustrated in Fig. 2. In recent years, a lot of work attempt to reduce the amount of calculation for graph convolution. Among these approaches, Cheng et al. [2] simplify skeleton-based network with shift graph convolutional network. In this paper, we introduce Shift-GCN to a spatial attention graph convolution module. What's more, we also use the time shift to simplify our model further.

2.3 Prototype Generation Module

By directly modeling the data or establishing a conditional probability distribution, the generative model can accurately predict the possible distribution of the data. Nowadays, a lot of work based on deep learning has introduced generative models. This trend is particularly evident in the field of unsupervised tasks.

For VAD, the generation module can effectively deal with the problem of 'over-generalizing'. Sabokrou et al. [7] believe that normal data obey an overall distribution and map them into a hypersphere. Yet they did not consider the complexity of the context and the diversity of normal patterns. Further, Gong et al. [8] and Park et al. [9] introduce a memory bank into the autoencoder for anomaly detection. Although they prove the performance of the methods, the memory cost has become a problem to consider. Recently, Lv et al. [3] propose to model the prototypes of normal data and use pixel-level attention mechanisms to locate normal patterns. We follow their Dynamic Prototype Unit (DPU), but we change the attention map to frames and human key points.

3 Method

In this section, we first give an overview of the dynamic prototype-based graph convolutional network (DP-GCN) and describe the details of the feature extracting module and prototype generation module later. Next, the details of the object function and implement will be introduced.

3.1 Architecture Overview

Our proposed DP-GCN is illustrated as Fig. 1, in which the feature extraction module obtains the representation of normal human pose sequence while the prototype generation module is designed to assign weights for feature maps and model the prototypes of normal data. Given a sequence of video frames, our method first exacts the human pose information. After a shift GCN autoencoder, the exacted features are sent to the prototype generation module. Finally, the aggregation features are re-clustered as the method mentioned in [1].

3.2 Feature Extraction Module

The based Spatio-temporal Graph Convolution Block (ST-GCN) is illustrated in Fig. 2. The spatial attention and shift GCN operation [2] is introduced into ST-GCN, and they construct a Shift Spatio-temporal GCN Auto Encoder (SST-GCAE).

As for the graph convolution block, the two branches of Spatial Attention GCN in [1] are retained, which use a globally-learnable matrix A and inferred adjacency matrices B. Another shift branch is utilized to simplify the model while ensuring the performance of graph convolution. The Shift GCN Spatial Attention block is illustrated in Fig. 3. $F \in \mathbb{R}^{C \times T \times V}$ represents the processed skeleton graph information, where the C, T, and V are the input dimension of the channel, time, and human key points. Three copies are generated and are performed standard graph convolution operations on two of them

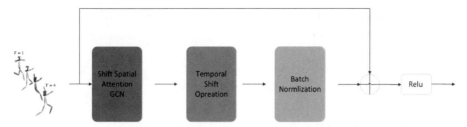

Fig. 2. The basic block used for constructing Shift Spatio-temporal GCN Auto Encoder (SST-GCAE). The module consists of a shift spatial attention GCN, a temporal shift operation block, and a batch normalization block. The shift spatial attention GCN is designed to embed the features of adjacent key points. The temporal shift operation block performs a typical shift operation in the time dimension. The result after regularization is aggregated with the residual branch and sent to the Activation function.

with A and B. It should be noted that the size of B is $[N, V, V]$, while adjacency A is $[V, V]$ matrices branch. So that the branch of A can capture dataset-level key point relations and the branch of B can capture batch-level relations. The third copy of F is sent to perform shift operations. For each frame of feature map $F_t \in \mathbb{R}^{C \times V}$, the shift distance of $i - th$ channel is set to $i \bmod V$. Finally, the output of three branches are stacked in the channel dimension.

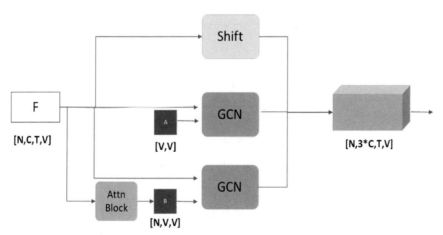

Fig. 3. The detail of Shift Spatial Attention Graph Convolutional Network. The N, C, T, and V are the input dimension of the batch size, channel, time, and human key points. The shift branch uses the topology of the human body after shift operation. And the adjacency matrixes A and B learned global information and attention-based information separately. The result of three branches are stacked into a $[N, 3 * C, T, N]$ feature map.

Behind the Shift GCN Spatial Attention block, the module also introduce the shift operation into the temporal convolution. After the Shift GCN Spatial Attention block, the embedded feature map $G \in \mathbb{R}^{C' \times T \times V}$ is given, where $C' = 3 \times C$. The shift distance of the time dimension is set as a learnable variable $D_i, i = 1, 2, ..., C'$. And the variable is no longer limited to an integer, but a real number:

$$\tilde{G}_{(v,t,i)} = (1-\lambda) \cdot G_{(v,\lfloor t+D_i \rfloor,i)} + \lambda \cdot G_{(v,\lfloor t+D_i \rfloor+1,i)} \tag{1}$$

where $\lambda = D_i - \lfloor D_i \rfloor$. After our lightweight SST-GCAE, information-rich features in space and time dimensions can be obtained.

3.3 Prototype Generation Module

The prototype generation module (PGM) aims to maps the normal data into prototypes and update them during the training phase. Inspired by [3], a prototype generation module is introduced into our network. Different from previous work, the proposed approach focuses on key frames and important joint points. By introducing the attention mechanism in the temporal dimension and key point dimension, the normal patterns of important positions and time periods can be captured.

Figure 4 shows the prototype generation module. Concretely, with the input feature maps $X \in \mathbb{R}^{C' \times T \times V}$, N attention maps are set to assign normalcy weights to encoding vectors. Here, $W^n \in \mathbb{R}^{T \times V}$ denotes the $n - th$ attention map. Then, N prototypes are generated:

$$p^n = \sum_{i=1}^{T} \sum_{j=1}^{V} \frac{w_{ij}^n}{\sum_{i'=1}^{T} \sum_{j'=1}^{V} w_{i'j'}^n} x_{ij} \tag{2}$$

Where $w_{ij}^n \in W^n$. Next, prototypes are used to reconstruct a normalcy feature map:

$$\tilde{x}_{ij} = \sum_{n=1}^{N} \beta_{ij}^n \cdot p^n \tag{3}$$

Where $\beta_{ij}^n = \frac{x_{ij}p^n}{\sum_{n'=1}^{N} x_{ij}p^n}$ denotes the relevant score between the vector at $i - th$ frame and the $j - th$ key points and the $n - th$ prototype. Finally, the normalcy feature is aggregated with the original X and apply a deep cluster layer to calculate the anomaly score. The detail of the cluster layer is the same as [1]. Since the cluster layer is not the focus of our research, we will not introduce it here.

3.4 VAD Objective Functions

The training process is divided into two steps. Firstly, we pre-train an autoencoder to reconstruct features. Next, we fine-tune our model by fixing decoder parameters.

Pre-training. This phase aims to extract the feature of normal data with our SST-GCAE and PGM. The loss function is composed of a reconstruction loss L_{rec} and a generation loss L_{gen}. L_{rec} is the L_2 distance of the input and the reconstructed pose graph map. And L_{gen} formula is as:

$$L_{gen} = L_c + \lambda L_d \tag{4}$$

Where λ is the weight parameter. L_c is the L_2 loss of the input vector of PGM and the closest prototype:

$$L_c = \frac{1}{T \times V} \sum_{i'=1}^{T} \sum_{j'=1}^{V} \left\| x_{ij} - p^* \right\|_2,$$

$$s.t., * = \arg \max_{1 \le n \le N} \beta_{ij}^n \tag{5}$$

L_d is used to keep distance between prototypes, ensure the diversity of prototypes.

$$L_d = \frac{2}{N(N-1)} \sum_{n=1}^{N} \sum_{n'=1}^{N} [-\left\| p^n - p^{n'} \right\|_2 + \gamma] \tag{6}$$

Where γ is set to control the margin between prototypes.

Fine-tuning. During this phase, the cluster loss $L_{cluster}$ is considered. Finally, the objective function is generated by combining three losses mentioned above:

$$L = L_{rec} + \lambda_1 L_{gen} + \lambda_2 L_{cluster} \tag{7}$$

Where λ_1 and λ_2 are the coefficient used to control the proportion of the three losses.

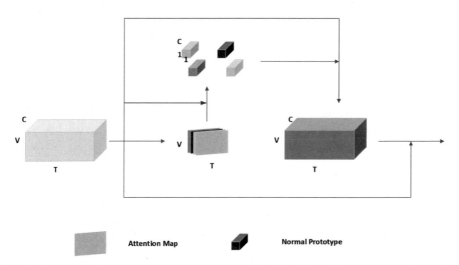

Fig. 4. The construction of prototype generation module. The C, T, and V are the input dimension of the channel, time, and human key points. The module firstly constructs **N** attention maps. In the figure, **N** is set to 4, and Generate **N** normal prototypes according to the weight of each attention map. Next, the input is split at the pixel level and the split vector is reconstructed according to the relationship with each prototype. After the prototype generation module, the feature map focus on crucial joint points and frames.

4   Experiments

We evaluated the performance of our model on the ShanghaiTech dataset [5] and an Ablation study is designed to verify the impact of each part of the model on the network.

4.1   Experiments on ShanghaiTech

Dataset. The ShanghaiTech dataset is one of the most commonly used databases in anomaly detection, which contains 330 training videos and 107 testing videos.The length of each video varies from 15 s to 1 min. The data is collected from 13 scenes. For each video clips, the number of people range from 0–20. All of the training video are normal video, and there are various of anomaly event in testing video like running, robbing and fighting.

Evaluation Metrics. Following previous work [4, 9, 10], we adopt Area under ROC Curve (AUC) as the indicator of model performance. By setting different thresholds and frame-level anomaly scores, it is easy to find the influence of any threshold on the generalization performance of the learner.

Implementation Details. We extract the human pose from the input RGB images and for each exacted pose, we expand 18 coordinate points outward into $18 \times 18 \times 3$ patches. Then we merge the patch information into the channel dimension, the channel is set to 64. We use 9 GCN layers to extract feature maps and add one PGM at the last layer. During the pre-training phase, the model is trained with the learning rate as 0.0008 and batch size as 64. The training epoch is set to 10. During the fine-tuning phase, we set the learning rate and batch size as 0.0001 and 128. The training epoch is set to 25. The balance weights in the objective functions are set as $\lambda_1 = 0.0001$, $\lambda_2 = 0.0001$, $\lambda = 1$. The experiment was carried out on a Nvidia RTX-2080Ti GPU.

Results. The AUC result on ShanghaiTech is shown in Table 1. Compared with other state-of-the-art methods, our DP-GCN achieves 76.7% AUC score, which outperforms the previous state-of-the-art semi-supervised methods by a large margin. When compared with weakly-supervised methods, our model also shows encouraging performance. Notice that our method doesn't need any annotation during the training phase, which has more extensive applications in real-world scenarios. [1] and [3] are the most related methods to our approach, it can be seen that our mothed outperforms them by 0.6 point and 2.9 points. The results prove that the combination of the generated prototype model and the shift graph convolutional network has excellent advantages.

Table 1. Experiments performance on ShanghaiTech

Supervision	Method	AUC
Semi-Supervised	Conv-AE [16]	60.9
	Stacked-RNN [5]	68.0
	Mem [4]	70.5
	Mem-AE [8]	71.2
	CGAN [6]	72.8
	U-Net [12]	73.0
	ST-GCN [1]	76.1
	DPU [3]	73.8
	rGAN*[11]	73.7
Weakly Supervised	MPED-RNN [13]	73.4
	Noisy labels + GCN [14]	76.4
	Deep MIL [15]	82.3
Semi-Supervised	Ours	**76.7**

4.2 Ablation Study

In this section, we first demonstrate the importance of each module in the network, and then, we analyze the impact of batch size and patch size. As shown in Table 2, the basic model comes from [1], in which the GCN module adopts 3 adjacency matrixes, and temporal convolution block is ordinary 2D convolution.

Shift Operation. As can be seen from the first two lines that the temporal shift is better than shift GCN operation, in the case of introducing only one-shift operation. The reason may be that the original temporal convolution cannot well extract the distribution feature of normal data in the time dimension. While the spatial attention graph convolution is able to extract spatial features well, due to its complex structure. Hence, the introduction of shift GCN makes the final result lower than the basic model. The result of the third line shows that the net result of combining the two shifts of temporal and spatial is

Table 2. Results of ablation experiments on ShanghaiTech Dataset

Shift GCN	Shift TCN	PGM	AUC
			76.1
√			75.5
	√		76.2
√	√		76.4
√	√	√	**76.7**

Table 3. Comparisons between different Batch Size and Patch Size

Batch Size	Patch Size	AUC
64	16	73.0
	18	76.4
	20	73.2
128	16	76.3
	18	**76.7**

better than using one of the networks alone. We finally add the PGM to our network, the performance of our network reach 76.7 AUC.

Patch Size. We tried to use different batch sizes and patch sizes. The result is shown in Table 3. In the case of the same batch size, the performance of the model is not positively correlated with the size of the patch size. On the one hand, the small size of patches causes the loss of scene information. On the other hand, setting the size too large will introduce unnecessary noise. When we set the patch size to 18, the captured image patches just cover the person.

Batch Size. We have chosen two batch size settings of 64 and 128. The performance of the latter is better than the former in general. The reason is that large batches help the model better learn the distribution of normal data.

5 Conclusion

In this paper, we proposed a novel Dynamic Prototype-based Graph convolutional network (DPM-GCN) for video anomaly detection. By replacing the basic graph convolutional with shift operation, DPM-GCN outperforms the original spatial attention graph convolution with less calculation. In addition, we introduced a prototype generation module to better predict the distribution of normal data. Sufficient Experiments prove that our method has superior performance compared with mainstream methods. Future work includes validate our model on more data sets and combine more scene information to optimize the model.

References

1. Markovitz, A., Sharir, G., Friedman, I., Zelnik-Manor, L., Avidan, S.: Graph embedded pose clustering for anomaly detection. In: Proceedings of the IEEE/CVF Conference on Computer Vision and Pattern Recognition, pp. 10539–10547 (2020)
2. Cheng, K., Zhang, Y., He, X., Chen, W., Cheng, J., Lu, H.: Skeleton-based action recognition with shift graph convolutional network. In: Proceedings of the IEEE/CVF Conference on Computer Vision and Pattern Recognition, pp. 183–192 (2020)

3. Lv, H., Chen, C., Cui, Z., Xu, C., Li, Y., Yang, J.: Learning normal dynamics in videos with meta prototype network. In: Proceedings of the IEEE/CVF Conference on Computer Vision and Pattern Recognition, pp. 15425–15434 (2021)
4. Park, H., Noh, J., Ham, B.: Learning memory-guided normality for anomaly detection. In: Proceedings of the IEEE/CVF Conference on Computer Vision and Pattern Recognition, pp. 14372–14381 (2020)
5. Luo, W., Liu, W., Gao, S.: A revisit of sparse coding based anomaly detection in stacked RNN framework. In: Proceedings of the IEEE International Conference on Computer Vision, pp. 341–349 (2017)
6. Liu, W., Luo, W., Lian, D., Gao, S.: Future frame prediction for anomaly detection–a new baseline. In: Proceedings of the IEEE conference on computer vision and pattern recognition, pp. 6536–6545 (2018)
7. Ruff, L., et al.: Deep one-class classification. In: International Conference on Machine Learning, pp. 4393–4402. PMLR (July 2018)
8. Gong, D., et al.: Memorizing normality to detect anomaly: Memory-augmented deep autoencoder for unsupervised anomaly detection. In: Proceedings of the IEEE/CVF International Conference on Computer Vision, pp. 1705–1714 (2019)
9. Luo, W., Liu, W., Gao, S.: Remembering history with convolutional lstm for anomaly detection. In: 2017 IEEE International Conference on Multimedia and Expo (ICME), pp. 439–444. IEEE (July 2017)
10. Mahadevan, V., Li, W., Bhalodia, V., Vasconcelos, N.: Anomaly detection in crowded scenes. In: 2010 IEEE Computer Society Conference on Computer Vision and Pattern Recognition, pp. 1975–1981. IEEE (June 2010)
11. Lu Y., Yu, F., Reddy M.K.K., Wang, Y.: Few-shot scene-adaptive anomaly detection. In: Vedaldi, A., Bischof, H., Brox, T., Frahm, J.M. (eds) Computer Vision – ECCV 2020. ECCV 2020. Lecture Notes in Computer Science, vol. 12350. Springer, Cham (2020). https://doi.org/10.1007/978-3-030-58558-7_8
12. Tang, Y., Zhao, L., Zhang, S., Gong, C., Li, G., Yang, J.: Integrating prediction and reconstruction for anomaly detection. Pattern Recogn. Lett. **129**, 123–130 (2020)
13. Morais, R., Le, V., Tran, T., Saha, B., Mansour, M., Venkatesh, S.: Learning regularity in skeleton trajectories for anomaly detection in videos. In: Proceedings of the IEEE/CVF Conference on Computer Vision and Pattern Recognition, pp. 11996–12004 (2019)
14. Zhong, J.X., Li, N., Kong, W., Liu, S., Li, T.H., Li, G.: Graph convolutional label noise cleaner: train a plug-and-play action classifier for anomaly detection. In: Proceedings of the IEEE/CVF Conference on Computer Vision and Pattern Recognition, pp. 1237–1246 (2019)
15. Sultani, W., Chen, C., Shah, M.: Real-world anomaly detection in surveillance videos. In: Proceedings of the IEEE Conference on Computer Vision and Pattern Recognition, pp. 6479–6488 (2018)
16. Hasan, M., Choi, J., Neumann, J., Roy-Chowdhury, A.K., Davis, L.S.: Learning temporal regularity in video sequences. In: Proceedings of the IEEE Conference on Computer Vision and Pattern Recognition, pp. 733–742 (2016)
17. Kipf, T.N., Welling, M.: Semi-supervised classification with graph convolutional networks. arXiv preprint arXiv:1609.02907 (2016)
18. Yan, S., Xiong, Y., Lin, D.: Spatial temporal graph convolutional networks for skeleton-based action recognition. In: Thirty-second AAAI Conference on Artificial Intelligence (April 2018)
19. Yu, B., Yin, H., Zhu, Z.: Spatio-temporal graph convolutional networks: A deep learning framework for traffic forecasting. arXiv preprint arXiv:1709.04875 (2017)
20. Zhao, J., Li, J., Cheng, Y., Sim, T., Yan, S., Feng, J.: Understanding humans in crowded scenes: deep nested adversarial learning and a new benchmark for multi-human parsing. In: Proceedings of the 26th ACM International Conference on Multimedia, pp. 792–800 (October 2018)

Computer Graphics

A New Image Super-Resolution Reconstruction Algorithm Based on Hybrid Diffusion Model

Jimin Yu[1], Jiajun Yin[1], Saiao Huang[1], Maowei Qin[1], Xiankun Yang[1], and Shangbo Zhou[2(✉)]

[1] College of Automation, Chongqing University of Posts
and Telecommunications, Chongqing, China
yujm@cqupt.edu.cn
[2] College of Computer Science, Chongqing University, Chongqing, China
shbzhou@cqu.edu.cn

Abstract. In the process of image denoising based on anisotropic diffusion model, the problem of edge information loss and "staircase effect" often appear. On the basis of anisotropic diffusion model, this paper combines the fractional diffusion model with the gradient based integer diffusion model, and introduces washout filter as the control term of the model, a new image super-resolution reconstruction algorithm based on hybrid diffusion model is proposed. In our proposed model, the fractional derivative will adjust its size adaptively according to the local variance of the image, and because the threshold k in the traditional diffusion function requires a lot of data experiments to get the best results, we also propose an adaptive threshold k function, whose value changes adaptively with the gradient of the image. Simulation results show that, compared with other algorithms, the new model still has a strong ability to retain image details and edge information after image reconstruction, and the introduced washout filter will also speed up the rapid convergence of the system to a stable state, and improve the convergence speed and stability of the system.

Keywords: Anisotropic diffusion · Fractional-order partial derivative · Mixed diffusion · Image super-resolution reconstruction

1 Introduction

Image resolution represents the number of pixels contained in a unit area of an image, and is one of the important indicators for evaluating image quality. The quality of the image will also directly affect the visual perception. The richness of the information contained in the image is closely related to the resolution of the image. The higher the image resolution, the image will convey more detailed information, and the visual perception effect of people will be better [1]. However, limited by image hardware equipment, environment, cost and other factors, in many fields such as security monitoring, medical treatment, low-resolution images can no longer meet people's needs. At this time, higher-resolution images are needed. The technology of restoring and reconstructing

© Springer Nature Singapore Pte Ltd. 2021
Y. Wang and W. Song (Eds.): IGTA 2021, CCIS 1480, pp. 173–188, 2021.
https://doi.org/10.1007/978-981-16-7189-0_14

a high-resolution image from a low-resolution image is essentially an ill-conditioned inverse problem. It has always been a research difficulty in the field of image processing. It can be roughly divided into two recovery methods, hardware and software. To improve the resolution of an image from the hardware aspect, a high-definition camera is often required, and the equipment investment and cost are relatively high. Therefore, the image-based super-resolution reconstruction technology has emerged as the times require. Image super-resolution reconstruction technology refers to the use of signal processing technology [2] on the basis of not improving the hardware conditions, to supplement the missing part of the input low-resolution image with high-value information, and perform processing on the input low-resolution image. The method to increase the resolution.

The super-resolution reconstruction technology based on a single image can be divided into three categories [3]: super-resolution technology based on interpolation, super-resolution technology based on reconstruction, and super-resolution technology based on learning. The interpolation-based image super-resolution reconstruction algorithm generally obtains the gray value of the pixel to be interpolated by weighting and estimating the pixel to be interpolated through the neighboring known pixels. The reconstruction-based super-resolution technology [4] is a method to establish different a priori models based on the information in the image degradation process, and to solve them in reverse to improve the image resolution. At present, most research is based on the image super-resolution reconstruction technology based on learning. Unlike other methods, the learning-based method includes a training step. A large number of high-resolution images will be used to construct a learning library to generate a learning model. In the process of image restoration, the prior knowledge obtained by the learning model is introduced to obtain the high-frequency details of the image and obtain a better image restoration effect. This article uses super-resolution technology based on reconstruction.

In recent years, image processing technology based on partial differential equations has developed rapidly, and it has good effects in image denoising and image super-resolution reconstruction. Afraites [5] proposed an image denoising model constrained and optimized by high-order partial differential equations. The partial differential equations are composed of second- order and fourth-order diffusion tensors, which combines the diffusion advantages of the PM model in uniform areas and near sharp edges, an efficient image denoising model is obtained. Tian [6] proposed two image segmentation and image denoising models based on partial differential equations, and constructed a trend fidelity term that can effectively suppress the staircase effect. At the same time, based on the wavelet transform image denoising model of curvature change, the level-enhanced image set is used to establish a curvature driving function based on the curvature of the level set, and then the curvature driving function is introduced as a correction factor into the variational model. The experiment verifies two improved models both have obvious denoising effects and good visual effects. Faiba [7] proposed a new hybrid image denoising algorithm based on integer order and fractional total variation. The proposed model is a combination of ROF model and fractional total variation, using the two models Advantages, and after introducing an appropriate norm space, it proves the existence and uniqueness of the model. The anisotropic diffusion method based on partial differential equation is a physical model-based image denoising algorithm. This method can obtain a better denoising effect after multiple iterations of the denoised image. Chen [8] proposed an image denoising model through adaptively weighted anisotropic diffusion, which combines local diffusion components and patch-based diffusion components to perform image denoising. In addition, a variable time step is also designed to solve the problem of excessive smoothing. Bai [9] proposed a generalized anisotropic diffusion image denoising model. The main idea is to introduce the G-derivative of the image intensity function into the general anisotropic diffusion model and use it for image denoising. Yin [10] proposed an image super-resolution reconstruction algorithm based on differential curvature-driven fractional nonlinear diffusion. Using a new edge indicator can better identify edges, slopes and flat areas.

Aiming at the problems of incomplete denoising, loss of high-frequency information and edge information, and ladder effect in the process of image denoising and super-resolution reconstruction based on anisotropic diffusion equation, and considering that the essence of fractional order is based on image nonlocal operator, it is introduced into image reconstruction, it can deal with the ladder effect of image better. In this paper, a new image super-resolution reconstruction algorithm based on mixed diffusion model is proposed by combining fractional order anisotropic diffusion model with integer order gradient diffusion model and introducing washout filter as the control term of the system. In the proposed new model, the fractional derivative will adjust its size adaptively according to the local variance of the image, and because the traditional diffusion function needs a lot of data to get the best results, we also propose an adaptive threshold K function, whose value changes adaptively with the gradient of the image.

2 Related Work

2.1 Characteristics of Fractional Differential Operators

The concept of fractional calculus can be traced back more than 300 years ago, was founded by Leibniz and Newton, and can be seen as a generalization of integer order. The research of fractional calculus mainly focuses on the analysis of time domain and frequency domain. Fourier transform and Laplace transform are two commonly used definitions of frequency domain. According to the differential properties of the Fourier transform, for any function $f(x, y) \in L^2(R)$, the following definition of the fractional partial derivative in the Fourier transform domain can be obtained:

$$D_x^v f(x, y) = F^{-1}\Big((jw_1)^v \hat{f}(w_1, w_2)\Big), \tag{1}$$

$$D_y^v f(x, y) = F^{-1}\Big((jw_2)^v \hat{f}(w_1, w_2)\Big), \tag{2}$$

where F^{-1} is a two-dimensional continuous Fourier transform operator, which represents the inverse Fourier transform.

As shown in Fig. 1, they are the amplitude-frequency characteristic curves of the fractional differential operator when the order of the fractional order is $v = 0.2, v = 0.5$,

$v = 0.7, v = 1, v = 2$. The frequency response of the fractional differential operator can be regarded as a nonlinear high-pass filter. When $\alpha > 0$, the high-frequency information of the signal will be nonlinearly enhanced, and the low-frequency information of the signal will be modified. That is to say, in areas where $\omega > 1$, such as the detailed texture part and edge part of the image, the enhancement effect of the fractional differential operator on its signal will be nonlinearly enhanced as the fractional order increases; at $0 < \omega < 1$ is the smooth area of the image, the suppression ability of the fractional differential operator is relatively weaker than that of the integer order, so it can better keep the amplitude of the smooth area unchanged. On the whole, the use of fractional differential operators can increase the amplitude of high-frequency information, while the intermediate and low-frequency information will be non-linearly preserved.

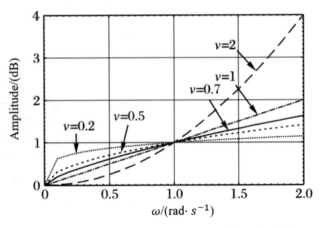

Fig. 1. Amplitude frequency characteristic curve of fractional differential

2.2 Fractional Anisotropic Diffusion Model

The anisotropic diffusion model was first applied to image denoising. Perona and Malik proposed a classic image denoising model based on the anisotropic diffusion equation in response to the shortcoming of heat conduction equations that cannot be selectively diffused in the denoising process [11] (PM diffusion model), its expression is:

$$\frac{\partial u}{\partial t} = div(C(|\nabla u|)\nabla u), \tag{3}$$

where div is the divergence symbol, $C(\cdot)$ is the diffusion coefficient, and ∇u represents the gradient of the image. From a mathematical point of view, the PM model uses the image gradient ∇u as an effective means to detect the flat area and the texture area of the image. In areas with large gradients such as the edge area and texture area of the image, the gradient of the image $|\nabla u|$ is relatively large, and a smaller diffusion coefficient will be used; in the non-edge area of the image, such as the flat area, the gradient of the image

$|\nabla u|$. Relatively small, a larger diffusion coefficient will be used. Therefore, Perona and Malik gave two diffusion functions $C(|\nabla u|)$ related to $|\nabla u|$ in their model.

$$C(\nabla u) = exp\left(-\left(\frac{|\nabla u|}{K}\right)^2\right),$$ (4)

$$C(\nabla u) = \frac{1}{\left(1 + \left(\frac{|\nabla u|}{K}\right)^2\right)}.$$ (5)

In Eqs. (4) and (5), k is a given parameter that controls the degree of diffusion. In this way, for the texture area of the image, the degree of diffusion of the function is small, thereby retaining the texture detail information of the original image, and for the noise part of the image, the degree of diffusion of the function is increased, so that the noise of the image can be effectively removed. However, with this diffusion method, the preservation of details in the image is not clear enough, and the image after image denoising and reconstruction is prone to have a staircase effect. Therefore, Bai [9] extended the gradient diffusion model based on integer order to fractional order, and proposed a class of image denoising model based on fractional anisotropic diffusion.

$$\frac{\partial u}{\partial t} = div(C(|\nabla^\alpha u|)\nabla^\alpha u)$$ (6)

The fractional diffusion function $C(|\nabla^\alpha u|)$ is used for edge detection, which can control the degree of diffusion according to the local characteristics of the image. Therefore, many models have appeared. The diffusion function is optimized to increase the accuracy of edge detection. Improve the denoising effect of the image. Zhang [12] realized a new method for calculating the spread function using the hyperbolic tangent function. V.B.S Prasath [13] proposed a fuzzy diffusion coefficient, which takes into account the variability of local pixels in order to better remove edge noise and selectively smooth the image.

The energy functional corresponding to Eq. (6) is:

$$E(u) = \int_\Omega f(|D_\alpha u|)d\Omega$$ (7)

Therefore, the denoising process of the image will be transformed into solving the minimum value problem of the energy functional. Then use the variational method of the functional to solve the Euler-Lagrange equation corresponding to the energy functional, and finally introduce the time variable t, and use the gradient descent method to transform into the following partial differential equation solution.

$$\frac{\partial u}{\partial t} = -D^*_{\alpha x}\left(c\left(|D_\alpha u|^2\right)D_{\alpha x}u\right) - D^*_{\alpha y}\left(c\left(|D_\alpha u|^2\right)D_{\alpha y}u\right)$$ (8)

3 Model Algorithm and Design

3.1 Proposal of Mixed Diffusion Model

In order to solve the problems of incomplete denoising, unclear image edges, and staircase effects in the reconstruction process when image denoising based on the fractional diffusion model at this stage, based on the anisotropic diffusion model, Combining the fractional-order diffusion model with the gradient-based integer-order diffusion model, and introducing the washout filter as the control item of the model, a new image super-resolution reconstruction and image denoising algorithm based on the hybrid diffusion model is proposed. In our new model, the fractional differential will adaptively change with the size of the local variance of the image. At the same time, for the diffusion functions $C(|\nabla u|)$ and $C(|\nabla^\alpha u|)$, we propose an adaptive threshold k function whose value changes adaptively with the gradient of the image. Combining the integer-order diffusion model and the fractional-order anisotropic diffusion model, the following model can be obtained.

$$E(u) = \int_\Omega f\left(|D^\alpha u|\right)d\Omega + \lambda \cdot \int_\Omega f(|\nabla u|)d\Omega. \tag{9}$$

Among them, λ is a parameter that balances integer-order diffusion and fractional-order diffusion, $|D_\alpha u| = \sqrt{D_{\alpha x}^2 + D_{\alpha y}^2}$, $|\nabla u| = \sqrt{D_x^2 + D_y^2}$ when $\alpha = 1$, it can degenerate to the classical PM model. In order to solve Eq. (9), we define the function:

$$h(\varepsilon) = E(u + \varepsilon\eta) = \int_\Omega f(|D_\alpha(u + \varepsilon\eta)|)d\Omega + \lambda \cdot f(|\nabla(u + \varepsilon\eta)|)d\Omega \tag{10}$$

Let $h'(\varepsilon)|_{\varepsilon=0} = 0$, we can get:

$$h'(\varepsilon)|_{\varepsilon=0} = \int_\Omega \left(D_{\alpha x}^* \left(c\left(|D_\alpha u|^2\right)D_{\alpha x} u \right) + D_{\alpha y}^* \left(c\left(|D_\alpha u|^2\right)D_{\alpha y} u \right) + \lambda \cdot \left(D_x^* \left(c\left(|Du|^2\right)D_x u \right) \right.\right.$$
$$\left.\left. + D_y^* \left(c\left(|Du|^2\right)D_y u \right) \right) \right) \eta \, dx \, dy. \tag{11}$$

Then further simplification can obtain the corresponding Euler-Lagrange equation.

$$D_{\alpha x}^* \left(c\left(|D_\alpha u|^2\right)D_{\alpha x} u \right) + D_{\alpha y}^* \left(c\left(|D_\alpha u|^2\right)D_{\alpha y} u \right) + \lambda \cdot \left(D_x^* \left(c\left(|Du|^2\right)D_x u \right) + D_y^* \left(c\left(|Du|^2\right)D_y u \right) \right) = 0. \tag{12}$$

Then, by using gradient descent method and introducing time variable t, the above Eq. (12) can be converted into the following nonlinear fractional partial differential equation.

$$\frac{\partial u}{\partial t} = -D_{\alpha x}^* \left(c\left(|D_\alpha u|^2\right)D_{\alpha x} u \right) - D_{\alpha y}^* \left(c\left(|D_\alpha u|^2\right)D_{\alpha y} u \right) - \lambda \cdot \left(D_x^* \left(c\left(|Du|^2\right)D_x u \right) \right.$$
$$\left. + D_y^* \left(c\left(|Du|^2\right)D_y u \right) \right) \tag{13}$$

Next, the washout filter is added to the nonlinear fractional differential equation, and the new model is obtained.

$$\begin{cases} \frac{\partial u}{\partial t} = -D_{\alpha x}^*\left(c\left(|D_\alpha u|^2\right)D_{\alpha x}u\right) - D_{\alpha y}^*\left(c\left(|D_\alpha u|^2\right)D_{\alpha y}u\right) - \lambda \cdot \left(D_x^*\left(c\left(|Du|^2\right)D_x u\right) + D_y^*\left(c\left(|Du|^2\right)D_y u\right)\right) + \beta\frac{\partial w}{\partial t} \\ \frac{\partial w}{\partial t} = -\gamma w + (u - u_0) \end{cases}$$

$$(14)$$

where, $\beta\frac{\partial w}{\partial t}$ is washout filter, which is the control term of the model, β. It is a weight control constant, which is used to balance the proportion between the diffusion term and the control term, and to maximize the effect of the diffusion term and the control term. Usually γ is a constant, and $\gamma = m/n$, where m, n are positive integers and satisfy $(m, n) = 1$.

3.2 Realization of Adaptive Diffusion Function

The two edge diffusion functions proposed in the classical PM model often have staircase effect in the process of image denoising. Therefore, a new adaptive diffusion function is constructed based on the edge spread function proposed by Guo.

$$c(|\nabla u|) = \frac{1}{1 + (|\nabla u|/K)^{\delta(|\nabla I|)}}, \qquad (15)$$

where $\delta(|\nabla I|) = 2 - \frac{2}{1+(|\nabla u|/K)^2}$, this diffusion function starts to weaken the diffusion from the place where the gradient is relatively large, which can effectively remove the noise and retain the texture details of the image to the maximum extent. There is also a threshold parameter k in the proposed diffusion coefficient, whose value determines the gradient value that begins to weaken the diffusion ability, which will affect the overall smoothing effect. If the value of k is too large, it will cause the image flat area to be too smooth and the image to be blurred; If value of k is too small, the image diffusion time will be shortened, and the denoising effect is not complete. Therefore, choosing a reasonable and accurate threshold parameter k will have a great impact on the overall effect of the model. Therefore, we propose an adaptive function of k, which will change adaptively with the gradient of the image. The expression of the threshold parameter k function is as follows.

$$k = k_0 e^{-\frac{1}{6}n\Delta t}, \qquad (16)$$

where $k_0 = mean(|\nabla u_0|)$, $mean$ is the average operator, n is the number of iterations, and Δt is the time step. The function of k is a decreasing function with the increase of the number of iterations. Its basic idea is: with the increase of the number of iterations, the peak signal ratio (PSNR) of the system will be larger and larger, the noise image will be smoother and smoother, the gradient of the image will be smaller and the value of k will be smaller.

At the same time, for the fractional diffusion model, the order of the fractional differential operator will adaptively change with the local variance of the image. The expression of the adaptive fractional α is shown below.

$$\alpha = 0.5 + exp\left(0.693 \times \frac{\sigma_{x,y} - min(\sigma_{x,y})}{max(\sigma_{x,y}) - min(\sigma_{x,y})}\right). \tag{17}$$

Among them, $\sigma_{x,y}$ is the variance of the image $u(x, y)$, $min(\sigma_{x,y})$ represents the minimum variance of the image, and $max(\sigma_{x,y})$ represents the maximum variance of the image. It can be seen from Eq. (17) that in the edge and other parts of the image, the local variance of the image is larger, so the order α is larger, and in the non-edge parts of the image, the local variance of the image is smaller, and the order α is smaller, and the more effective the noise can be removed.

3.3   Numerical Calculation of the Algorithm

In order to deal with the calculation of fractional derivative conveniently, we can first perform discrete Fourier transform on Eq. (14) to obtain its solution in frequency domain, and then obtain its solution in time domain through inverse Fourier transform. We will use the difference method to rewrite the left $\frac{\partial u}{\partial t}$ and $\frac{\partial w}{\partial t}$ in f Eq. (14) into the difference scheme $\frac{u_{k+1}-u_k}{\Delta t}$, $\frac{w_{k+1}-w_k}{\Delta t}$, where Δt is the time step, u_k is the previous output, which is used to calculate the next iteration result u_{k+1}. The following formula can be obtained by Fourier transform on both sides of the formula.

$$\frac{u_{k+1} - u_k}{\Delta t} = F\Big(-D_{\alpha x}^*\big(c(|D_\alpha u|^2)D_{\alpha x}u\big) - D_{\alpha y}^*\big(c(|D_\alpha u|^2)D_{\alpha y}u\big) - \lambda$$

$$\Big(D_x^*\big(c(|Du|^2)D_x u\big) + D_y^*\big(c(|Du|^2)D_y u\big)\Big)\Big) + \gamma \cdot \beta w_{k+1} + \beta \cdot (u_{k+1} - u_k)$$

$$= -K_1^* \times F\big(D_{\alpha x}^*(c(|D_\alpha u|^2)D_{\alpha x}u)\big) - K_2^* \times F\Big(D_{\alpha y}^*\big(c(|D_\alpha u|^2)D_{\alpha y}u\big)\Big) - \lambda \times K_3^*$$

$$\times F\big(D_x^*(c(|Du|^2)D_x u)\big) - \lambda \times K_4^* \times F\Big(D_y^*\big(c(|Du|^2)D_y u\big)\Big) + \gamma \cdot \beta w_{k+1} + \beta \cdot (u_{k+1} - u_k)$$

$$\frac{w_{k+1} - w_k}{\Delta t} = -\gamma w_{k+1} + (u_{k+1} - u_k). \tag{18}$$

Where F represents the Fourier transform, K_1^*, K_2^* respectively represent the conjugate operator of the α-order differential operator in the x direction and y direction of the image, and K_3^*, K_4^* is expressed as the conjugate operator of the first-order differential operator in the x and y directions of the image respectively.

$$K_1^* = conj(1 - exp(-j2\pi\omega_1/m))^\alpha \times exp(-j\pi\alpha\omega_1/m)$$
$$K_2^* = conj(1 - exp(-j2\pi\omega_2/m))^a \times exp(-j\pi\alpha\omega_2/m)$$
$$K_3^* = conj(1 - exp(-j2\pi\omega_1/m)) \times exp(-j\pi\alpha\omega_1/m)$$
$$K_4^* = conj(1 - exp(-j2\pi\omega_2/m)) \times exp(-j\pi\alpha\omega_2/m) \tag{19}$$

Therefore, the proposed image super-resolution reconstruction algorithm based on hybrid diffusion model is shown in Algorithm 1.

Algorithm 1: image super-resolution reconstruction based on hybrid diffusion model

Input: Initial low resolution image $u(x, y)$, time step $\Delta t = 0.01$, $\beta = 0.1$, $\gamma = 2/5$, and
          initialization $k = 1$.

Output: Image $\hat{u}(x, y)$ after super- resolution reconstruction.

do

    1.   Calculate the adaptive fractional order α of the image according to equation (17);

    2.   Solving D_{ax}^*, D_{ay}^*, D_x^*, D_y^*;

    3.   Calculate the adaptive spread function $c(|\nabla u|)$ according to equation (15);

    4.   Calculate $c(|D_\alpha u|^2)D_{ax}u$, $c(|D_\alpha u|^2)D_{ay}u$, $c(|Du|^2)D_x u$, $c(|Du|^2)D_y u$;

    5.   Solving K_1^*, K_2^*, K_3^*, K_4^*;

    6.   According to equation (18), iteratively calculate w_{k+1} and u_{k+1}, and set the number of
          iterations $k = k + 1$. When the PSNR value is maximum, the iteration is terminated and
          the reconstructed image is output.

End for

---

4 Experiment and Analysis

All experiments in this paper are run on a laptop computer configured as Inter (R) core (TM) i5-4200h CPU @ 2.80 GHz, and all programs are implemented by MATLAB R2018b. In the experiment, we choose six images such as "Lena", "butterfly" and "bird" as test images. In order to verify the effectiveness of the proposed super-resolution algorithm based on hybrid diffusion model, the test results of this algorithm are compared with the super-resolution reconstruction method of adaptive sparsity image [14] (ASDS), the super-resolution method of anchoring neighborhood regression [15] (A+),the super-resolution reconstruction method of image based on deep convolution network [16] (SRCNN) and the denoising algorithm based on generalized anisotropic diffusion equation [17] (GAD) are compared. Here, we take PSNR and SSIM as the main evaluation indexes of image super-resolution reconstruction. In the experiment, the test image is magnified twice and three times in sequence, and the corresponding PSNR values and SSIM values of the respective algorithms were compared, and the data as shown below were obtained.

Table 1. PSNR and SSIM comparison table of different methods at twice magnification.

Image	Scale	ASDS	A+	SRCNN	GAD	Proposed
Head	×2	27.1130	31.9346	32.9043	32.7293	**33.6415**
		0.7576	0.7717	0.7941	0.8089	**0.8685**
Lena	×2	29.7871	29.4512	33.5293	33.4886	**34.2691**
		0.8328	0.8755	0.8971	0.8814	**0.9321**
Butterfly	×2	21.5973	**28.3173**	26.5581	26.3546	27.3518
		0.7838	0.9159	0.9047	**0.9245**	0.9014
Bird	×2	28.5741	32.1194	34.9014	35.0074	**35.5666**
		0.8730	0.8903	0.9403	0.9621	**0.9713**
Baby	×2	31.0585	32.2132	33.2786	34.0296	**35.8913**
		0.8965	0.8754	0.8921	0.9427	**0.9489**
Peppers	×2	29.3634	29.9529	31.3301	31.1792	**31.8173**
		0.8117	0.8510	0.8631	**0.8723**	0.8465

Table 2. Comparison table of PSNR and SSIM for different methods when magnification is three times.

Image	Scale	ASDS	A+	SRCNN	GAD	Proposed
Head	×3	27.0995	27.8996	27.7842	27.8417	**30.3625**
		0.8035	0.7532	0.8165	0.8446	**0.8610**
Lena	×3	27.0728	28.3499	29.5133	30.1058	**30.7992**
		0.8043	0.7801	0.8044	0.7893	**0.8679**
Butterfly	×3	20.5918	23.4902	26.3010	**27.1006**	26.8146
		0.7232	0.7853	0.8118	0.8715	**0.8756**
Bird	×3	26.0868	27.8103	29.0654	28.6141	**29.4197**
		0.7991	0.8249	0.8609	0.9018	**0.9132**
Baby	×3	28.2968	29.9442	31.1893	30.6984	**32.5712**
		0.8326	0.8115	0.8440	0.7769	**0.9002**
Peppers	×3	26.8017	28.3244	29.1079	28.9743	**29.6572**
		0.7634	0.7910	0.8110	**0.8652**	0.8558

As can be seen from Table 1 and Table 2, the super-resolution images of six different images with different magnification are reconstructed with different reconstruction algorithms, and the bold data in the table is expressed as the maximum values of PSNR and SSIM obtained under the same conditions. It can be seen from the data in the table that the hybrid diffusion algorithm proposed in this paper is better than other reconstruction algorithms in terms of image quality and the integrity of image structure information and detail information. At the same time, as shown in Fig. 2 and Fig. 3, we select "Butterfly" and "bird" images for the contrast experiment of twice and triple magnification, which contains a lot of texture and structure information.

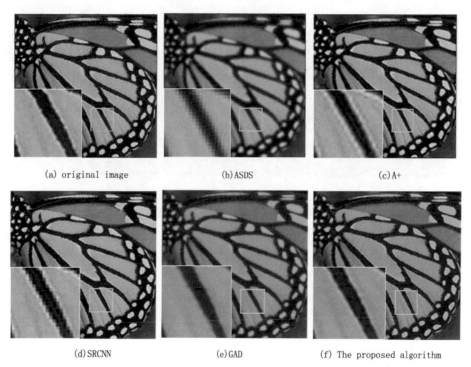

(a) original image   (b) ASDS   (c) A+

(d) SRCNN   (e) GAD   (f) The proposed algorithm

Fig. 2. Reconstruction results of "Butterfly" image by different algorithms.

As shown in Fig. 2 and Fig. 3, the reconstruction effects of different algorithms on "Butterfly" and "bird" images with different magnification are given respectively. At the same time, from the comparison of PSNR and SSIM values, it can be comprehensively verified that the image reconstruction effect of this algorithm is obviously better than other algorithms. It can be seen from the comparison figure that the images reconstructed by ASDS algorithm and A+ algorithm have obvious sawtooth effect and staircase effect; about based on SRCNN algorithm, the whole reconstructed image is too smooth, the local area has low contrast, and the edge area also has a certain staircase effect; Although the effect of the reconstructed image based on GAD algorithm is better than other algorithms, the phenomenon of local blur appears. After using the hybrid diffusion algorithm in this paper to reconstruct the image, the sawtooth effect and staircase effect of the image are

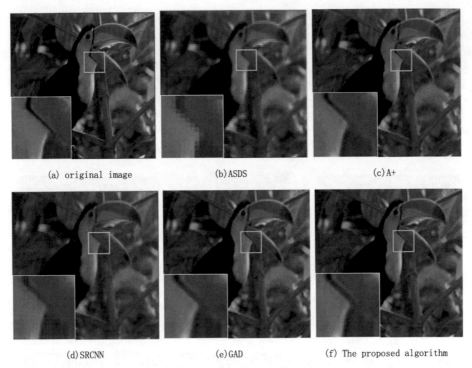

(a) original image   (b) ASDS   (c) A+

(d) SRCNN   (e) GAD   (f) The proposed algorithm

Fig. 3. Reconstruction results of "Bird" image by different algorithms.

significantly reduced, the edge smoothness of the image is better, and the reconstructed image is closer to the original image. Therefore, in general, this algorithm has more advantages than other traditional image reconstruction algorithms, and has a certain research value.

In order to more intuitively reflect the superiority of the proposed algorithm in image super-resolution reconstruction, we have drawn the line graphs between the PSNR and SSIM values of each algorithm and different test images, as shown in Fig. 4 and Fig. 5. The abscissa represents different test images, the ordinate represents the PSNR and SSIM values, and each colored broken line represents a different algorithm. It can be seen more intuitively from Fig. 4 and Fig. 5 that the algorithm proposed in this article has a higher value whether it is in the PSNR value or the SSIM value.

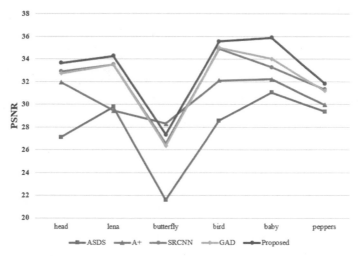

Fig. 4. PSNR of different algorithms for different images.

In order to examine the overall comprehensive performance of the algorithm in this article and the other four comparison algorithms, first, according to Table 1 and Table 2, we find the average PSNR value and average SSIM value of each algorithm on different images, and then according to the PSNR and SSIM average values Evaluate the overall performance of each algorithm. Therefore, we take the average PSNR value of each algorithm as the X axis and the average SSIM value as the Y axis, and draw the scatter diagram as shown in Fig. 6. The farther the algorithm's corresponding scatter points are from the origin of the coordinates, the overall performance will be the better. It can be seen intuitively from Fig. 6 that the overall performance of the ASDS algorithm is the worst, while the overall performance of the algorithm in this paper is the best.

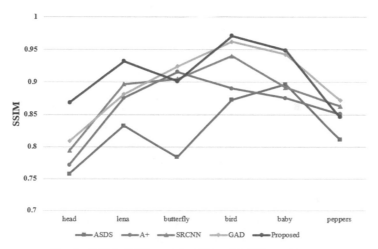

Fig. 5. SSIM of different algorithms for different images.

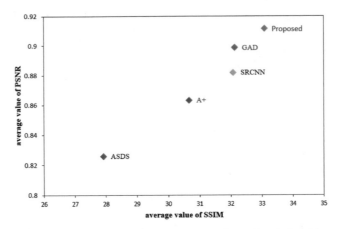

Fig. 6. Comprehensive performance comparison of each algorithm.

At the same time, the washout filter is added as the control term of the model in this algorithm. The simulation results show that the washout filter as the control term can accelerate the convergence speed of the system and improve the stability of the system. In the experiment, "bird" image is selected as the reference. Firstly, Gaussian noise with mean value of 0 and variance of 15 is added to the image, and then the image super-resolution reconstruction and image denoising simulation experiment are carried out. The growth relationship between the PSNR value in the experiment and the iteration number k of the system is compared.

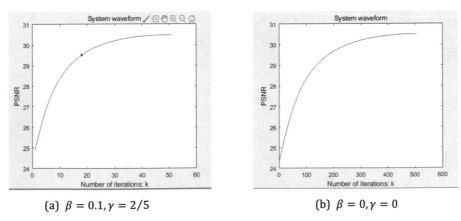

(a) $\beta = 0.1, \gamma = 2/5$ (b) $\beta = 0, \gamma = 0$

Fig. 7. Waveform of PSNR value and iteration number k.

As shown in Fig. 7 (a), the waveform of PSNR value and iteration number k when washout filter is introduced. When the iteration number k is about 40, the system tends to be stable; Fig. 4 (b) shows the waveform of PSNR value and iteration number k without washout filter. Even if the iteration number k is about 400, the system is still not stable. It can be seen from Fig. 4 that when washout filter is introduced into the model, the effect of fast convergence and stability of the system can be achieved.

5   Conclusion

The existing anisotropic diffusion equation can achieve good visual effect in image denoising and image reconstruction, but in the process of denoising and reconstruction, it is easy to appear the phenomenon of fuzzy edge details and staircase effect, which affects the subsequent image feature extraction, image decomposition and other image processing technologies. To solve this problem, this paper proposes an image super-resolution reconstruction algorithm based on hybrid diffusion model. Because the traditional diffusion function requires a large number of data experiments to get the best results, we propose an adaptive threshold k function, whose value changes adaptively with the image gradient. At the same time, by introducing washout filter as the control term of the system, the fast convergence of the system is achieved. Experimental results show that the proposed algorithm is superior to other traditional algorithms in many indexes.

References

1. Yang, X., Xie, T., Liu, L.: Image super-resolution reconstruction based on improved Dirac residual network. Multidimens. Syst. Signal Process. 1–18 (2021)
2. Zhao, X., Zhang, X.: Multi-frame super-resolution reconstruction algorithm of optical remote sensing images based on double regularization terms and unsupervised learning. Int. J. Pattern Recogn. Artif. Intell. **35**, 2154002 (2020)
3. Laghrib, A., Hadri, A., Hakim, A.: An edge preserving high-order PDE for multiframe image super-resolution. J. Franklin Inst. **356**, 5834–5857 (2019)
4. Khan, A., Khan, M.A.: A novel multi-frame super resolution algorithm for surveillance camera image reconstruction. In: First International Conference on Anti-cybercrime. IEEE (2015)
5. Afraites, L., Hadri, A., Laghrib, A.: A high order PDE-constrained optimization for the image denoising problem. Inverse Probl. Sci. Eng. 1–43 (2020)
6. Tian, C., Chen, Y.: Image segmentation and denoising algorithm based on partial differential equations. IEEE Sens. J. **20**, 11935–11942 (2020)
7. Golbaghi, F.K., Rezghi, M., Eslahchi, M.R.: A hybrid image denoising method based on integer and fractional-order total variation. Iran. J. Sci. Technol. Trans. A Sci. **44**(3), 1803–1814 (2020)
8. Chen, Y., He, T.: Image denoising via an adaptive weighted anisotropic diffusion. Multidimension. Syst. Signal Process. **32**(2), 651–669 (2021). https://doi.org/10.1007/s11045-020-00760-x
9. Bai, J., Feng, X.C.: Fractional-order anisotropic diffusion for image denoising. IEEE Trans. Image Process. **16**, 2492–2502 (2007)
10. Yin, X.: Fractional-order difference curvature-driven fractional anisotropic diffusion equation for image super-resolution. Int. J. Model. Simul. Sci. Comput. **10**(01), 1941012 (2019)
11. Malik, J.M., Perona, P.: Scale-space and edge detection using anisotropic diffusion. IEEE Trans. Pattern Anal. Mach. Intell. **12**(7), 629–639 (1990)
12. Zhang, Y., Sun, J.: An improved BM3D algorithm based on anisotropic diffusion equation. Math. Biosci. Eng. **17**(5), 4970–4989 (2020)

13. Surya Prasath, V.B., Delhibabu, R.: Image restoration with fuzzy coefficient driven anisotropic diffusion. In: Panigrahi, B.K., Suganthan, P.N., Das, S. (eds.) SEMCCO 2014. LNCS, vol. 8947, pp. 145–155. Springer, Cham (2015). https://doi.org/10.1007/978-3-319-20294-5_13
14. Dong, W., Zhang, L., Shi, G., Wu, X.: Image deblurring and super-resolution by adaptive sparse domain selection and adaptive regularization. IEEE Trans. Image Process. **20**(7), 1838–1857 (2011)
15. Timofte, R., De Smet, V., Van Gool, L.: A+: adjusted anchored neighborhood regression for fast super-resolution. In: Cremers, D., Reid, I., Saito, H., Yang, M.-H. (eds.) ACCV 2014. LNCS, vol. 9006, pp. 111–126. Springer, Cham (2015). https://doi.org/10.1007/978-3-319-16817-3_8
16. Dong, C., Loy, C.C., He, K.: Image super-resolution using deep convolutional network. IEEE Trans. Pattern Anal. Mach. Intell. **38**(2), 295–307 (2016)
17. Bai, J., Feng, X.-C.: Image denoising using generalized anisotropic diffusion. J. Math. Imaging Vis. **60**(7), 994–1007 (2018). https://doi.org/10.1007/s10851-018-0790-4

Research on Global Contrast Calculation Considering Color Differences

Jing Qian[1,2,3]([✉]) and Bin Kong[1,2,3]

[1] Institute of Intelligent Machines, Chinese Academy of Sciences, Hefei 230031, China
bkong@iim.ac.cn
[2] University of Science and Technology of China, Hefei 230026, China
qjjq@mail.ustc.edu.cn
[3] Peng Cheng Laboratory, Shenzhen 518053, China

Abstract. Image quality evaluation is an important research topic in the field of image processing. Contrast is a common image quality assessment index which can reflect the level of difference between the colors. However, the traditional method of global contrast calculation always has misdescribe when uniform color or nearly uniform color blocks appear in the image. We propose a global contrast method based on RGB-difference among regions. First, divide the image into several regions, and get the difference information of RGB-components between different regions. Second, calculate the parameter value of grayscale transformation which preserve the contrast information of the original image. Finally, combine the difference information and the parameter value of grayscale transformation to get the global contrast information. Compared with traditional method, our method is less affected by the uniform color blocks of the experimental image, it can describe the contrast of experimental image more objectively and fairly. In addition, the contrast value obtained by our method is consistent with the trend of human visual judgement, and conform to the objective law.

Keywords: Global contrast · RGB-difference · Grayscale transformation

1 Introduction

In recent years, with the development of technology, image processing technology is widely used in intelligent devices and become a hot research area [1, 2]. In actual projects, different methods and processes will affect the quality of the result image, and the quality of image directly affects the access of information and subjective feelings. Therefore, image quality assessment is a very important research topic in the field of image processing [3, 4].

In the common image quality assessment values [5, 6], global contrast reflects the level of difference between the colors or between bright and dark areas in an image, that is, the degree to which the details in the image can be distinguished [7, 8]. Global contrast is important in object detection and recognition [9, 10]. A high-contrast image has higher definition and clarity with more gray levels and details and identifying objects

© Springer Nature Singapore Pte Ltd. 2021
Y. Wang and W. Song (Eds.): IGTA 2021, CCIS 1480, pp. 189–200, 2021.
https://doi.org/10.1007/978-981-16-7189-0_15

in it would be easier, while a low-contrast image appears hazy and identifying objects in it would be difficult. Traditional calculation of global contrast depends on the absolute grayscale value of pixel rather than considering the color information, and do not take into account the color distribution characteristics between regions [11, 12]. However, when there are large uniform or nearly uniform color blocks in the image, the traditional contrast calculation value will be relatively smaller, this cannot reasonably and fairly represent the contrast information of original image.

Image contrast has a direct impact on visual observation results, its value should not only be reflected in the numerical difference between adjacent pixels, it should also be reflected in the color difference between different regions of the image [13, 14].

This paper proposes a method to calculate the global contrast value based on RGB-difference between different regions. Firstly, we calculate the difference of R, G and B channels in different regions of the image. Then get the linear decoloring parameters which retain the most information of image contrast. Finally, combine the difference information of R, G, B channel with the linear decoloring parameters to get the final contrast information. We believe that this work can provide a new idea for the calculation of global contrast.

2   Related Work

The global contrast is the most crucial characteristic of the image, which largely determines its visual perception. In this section, we briefly introduce global contrast measures defined in the literature.

The Weber-Fechner and Michelson contrasts are the first global contrast definitions which has been widely used in many scenes, such as contrast enhancement, image quality assessment, and quantization. These contrast methods are used to study the HVS (Human Visual System) mechanisms or physical phenomena in specific scene. Although these contrast methods are simple and limited to specific experiments, they have been used in many image analysis and processing methods [12, 15, 16].

Lillesaeter contrast is considered to be the first attempt to consider both photometric and geometric aspects in the computation of contrast information [17]. First, this method only considers the contrast definition of brightness, and then integrates the form of the perceived object [18, 19]. However, in practice, it is not easy to use, because it requires a priori knowledge of the object contour and computation of the curvilinear integral along the boundaries of the objects contained in the observed image.

In 2003, Calabria and Fairchild introduced two global measures for perceived contrast in color images in the CIELAB 1976 color space [20]. One is called Reproduction Versus Preferred (RVP) contrast and the other one is the Single Image Perceived (SIP) contrast. These two concepts are based on the observer perceptual contrast preference. The contrast model is established by studying the relative lightness, chroma, and sharpness. However, the key parameters of the experiment are only obtained by simple linear regression model experiments, this limits the reliability of the method.

Some methods based on statistical has been proposed for practical applications [21–23]. While these methods could not be fully considered as a contrast measures, they express or contain some information related to the contrast. These methods have only one single value associated with each image, and the calculation is ease and simple, they have been widely used in practical applications, from texture and image classification to face detection [10].

In 2017, Shaus et al. proposed a Potential Contrast (PC) method [24], this method is based on the Weber-Fechner's model and calculates the contrast information by dividing the image into the background and the foreground. Although the method conforms to the mathematical concepts and criteria set out by the authors, the method ignores the frequency and directional of the structures in images.

In 2020, Yuriy et al. proposed a method to quickly quantify the contrast of the image by measuring its incomplete integral contrast [8]. At the heart of this method is the assessment of contrast for each object in the image relative to the adaptation level given the sizes (area) of these objects. However, it is not easy to use in practice, because it requires a priori information of the object in the image.

3 Method

In this section, we introduce a widely used traditional method of global contrast calculation, and point out its error in specific environment. Then we describe the calculation process of contrast calculation method based on RGB-difference among regions.

3.1 Traditional Method of Global Contrast

The traditional method of global contrast calculation is to calculate gray difference between pixel and other pixels, so as to reflect the administrative levels and definition of images.

Haralick et al. proposed a global contrast method based on the Gray Level Co-occurrence Matrix (GLCM) computed from the luminance component of a digital image [21, 26]. This method can capture the average local variations and spatial dependence of the pixels. Because of simple calculation and easy expression, the calculation idea of GLCM has been proposed for various applications. The global contrast of GLCM can be expressed by formula (1):

$$C = \sum_{\delta} \delta(I, J)^2 P_{\delta}(I, J) \tag{1}$$

In formula (1), two adjacent pixel points are denoted by I and J, $\delta(I, J) = |I - J|$ is the value of gray difference between two adjacent pixel points, $P_{\delta}(I, J)$ is the probability of distribution when the gray difference between adjacent pixels is $\delta(I, J)$.

The traditional image grayscale processing algorithm is as Formula (2):

$$Gray = R * 0.299 + G * 0.587 + B * 0.114 \tag{2}$$

In generally, traditional method of GLCM can objectively reflect the contrast degree of the image. However, when there are large uniform or nearly uniform color blocks in the image, the value of traditional global contrast calculation will be different from the visual perception results of human beings (see Fig. 1).

(a) C=61.5   (b) C=88.7

Fig. 1. Results of traditional contrast. (Color figure online)

In Fig. 1, C is GLCM value of the image. The yellow object in the left image is obviously different from the background environment. And the texture details are clear, in contrast, there is obvious blurred in the right picture. From the perspective of subjective vision, we judge that the visual effect of the left image is better than that of the right image. However, due to the large similar area in the background of the left image, the color difference of this area is small, the GLCM value of left image is smaller than that of the right image. This is inconsistent with the judgment of human visual observation. In this case, the traditional global contrast method cannot reasonably and objectively reflect the visual effect of the image.

We design a special synthetic images (see Fig. 2) to analyze the problem of traditional global contrast value.

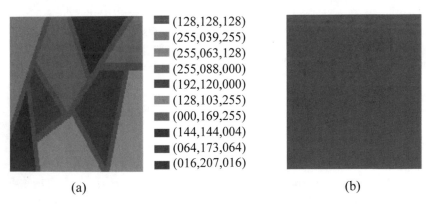

(a)   (b)

Fig. 2. An image (a) with different colors, but its gray values calculated by the formula (2) are exactly the same on all pixels (b). This causes the contrast value C = 0.

Figure 2(a) is an image composed of pure color regions. From the perspective of human vision, the image has obvious color differences between different regions, but its contrast value C is zero. This is because after transforming it from rgb to grayscale, the values of all pixels in it will be the same 128, which means the color difference is lost.

In order to objectively explain the problems of traditional global contrast calculation method. We make different combinations of pure color regions to obtain the experimental image, and display the contrast calculation results of the GLCM method with different RGB channel combinations. The different combinations are as follows:

- Use formula (2) to get the grayscale image, and then use the GLCM method to calculate the contrast C_1.
- First, calculate the GLCM values C_R, C_G, C_B corresponding to the R, G, and B channels of the image. Then contrast value $C_2 = C_R * 0.299 + C_G * 0.587 + C_B * 0.114$.
- Contrast value $C_3 = (C_R + C_G + C_B)/3$.
- Contrast value $C_4 = (C_1 + C_R * 0.299 + C_G * 0.587 + C_B * 0.114)/2$.

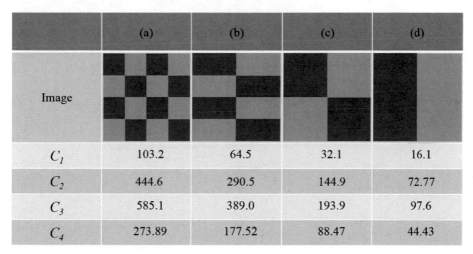

	(a)	(b)	(c)	(d)
Image				
C_1	103.2	64.5	32.1	16.1
C_2	444.6	290.5	144.9	72.77
C_3	585.1	389.0	193.9	97.6
C_4	273.89	177.52	88.47	44.43

Fig. 3. GLCM method results of different experimental images.

In Fig. 3, each image consists of sixteen square regions of the same size and pure color, the components are eight blue (RGB value is (0, 0,255)) color regions and eight green (RGB value is (0, 200,0)) color regions. Figure 3(a) image shows the 16 color regions staggered combination by color difference; Fig. 3(b) image shows different color regions staggered combination by color difference after two regions of the same color are combined; Fig. 3(c) image shows different color regions staggered combination by color difference after four regions of the same color are combined; Fig. 3(d) image shows color regions combination after the same color are combined. In Fig. 3, the number and value of pixels in the four images are the same, and the only difference is the arrangement of the color regions. From the perspective of human vision, the contrast perception of the four images are similar.

From the data in Fig. 3, we can find that GLCM values C are different. And the more the same color regions combination, the smaller the value of C. This result is in contradiction with the fact that the contrast perception of four images is consistent in human visual observation. It shows that the traditional global contrast calculation method has a large error in this case. Because the trend of each C value is consistent, we use the C_1 calculation method as the contrast value of the GLCM method in the following.

Contrast information is a kind of psychological feeling, it contains the whole image. The value of contrast should take into account the differences between different color regions of the image.

3.2   Global Contrast Calculation Considering Color Differences

In this paper, we propose a new method of global contrast computation by combine the difference information of RGB channels between different regions with the optimization results of the linear decoloring parameters. First, divide the original image is into several regions of equal size. Then calculate the average value of three channels in each region, and count the difference value of three channel average values among the regions, get the difference information of three channels in different regions. At the same time, calculate the linear decoloring parameter value with the image contrast information preserved. Finally, combine the difference information of three channels in different regions with the linear decoloring parameter to get the contrast information. The flow diagram of the proposed method is shown as Fig. 4 in the following subsections, we present the detailed description of the process of our method.

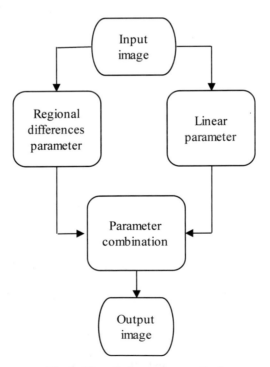

Fig. 4. Flow diagram of our method.

3.2.1 Color Difference Among Regions

First of all, we divide the original image into $m * n$ regions of equal size, among them, m is the average number of horizontal axis divided, n is the average number of vertical axis divided. And calculate the average value $A_{z(i,j)}$ of R, G and B channels in each region. $A_{z(i,j)}$ define the average value of Z-channel values in the region of (i, j), and $Z \in (R,G,B)$.

The difference information of three channels in different regions (denoted as D_Z) can be expressed by Eq. (3):

$$D_z \atop z \in (R,G,B) = \frac{\sum \Delta A_z^2 P_{\Delta A_z}}{n * m} \tag{3}$$

In Eq. (3), ΔA_z denote the difference value of Z-channel mean value between two regions, $P_{\Delta A_z}$ represents the probability value that the mean difference among regions is ΔA_z.

In order to combine the difference information into a contrast value which is consistent with the observation effect of human eyes, we need to combine the difference values of each channel in a scientific proportion.

3.2.2 Contrast Preserving Grayscale Processing

In most time, the linear combination ratio of R, G and B channels in the traditional image decoloring algorithm is consistent with the human visual observation effect.

The linear combination of image grayscale processing algorithm is as Formula (4):

$$g = R * w_r + G * w_g + B * w_b \tag{4}$$

And the restraint condition of (w_r, w_g, w_b) is Eq. (5):

$$w_r > 0; \; w_g > 0; \; w_b > 0 \\ w_r + w_g + w_b = 1 \tag{5}$$

The traditional grayscale transformation ratio is $w_r = 0.299$, $w_g = 0.587$, $w_b = 0.114$.

However, in some application scenes, the result of using the traditional decolor conversion ratio is not good, the difference of the different color in the original picture is smaller after transformed into a grayscale image, the original contrast information is lost. As shown in Fig. 6(a) and Fig. 6(b).

By comparing the two images of Fig. 5(a) and Fig. 5(b), some obvious color differences in the original picture, such as the yellow-blue difference at the edge, is difficult to detect after traditional grayscale processing. This loss of contrast information may not be a problem for general image processing, but it is very disadvantageous to analyze the accuracy of objective evaluation parameters of image.

Therefore, it is necessary to calculate the linear combination ratio of R, G and B channels to retain the color contrast information of the original image, this kind of

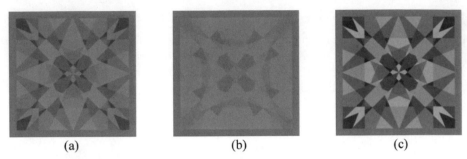

(a)   (b)   (c)

Fig. 5. Different grayscale results (a) Original image; (b) Grayscale result of formula (2); (c) Grayscale result with contrast information (Color figure online)

algorithm usually aims at minimizing the following energy function (6) to preserve the contrast information [27]:

$$\min_{g} \sum_{x,y} \left(g_x - g_y - \delta_{x,y}\right)^2 \tag{6}$$

In function (6), g_x and g_y is the gray value of point x and point y. δx, y denotes the color contrast of point x and point y. Based on the Euclidian distance in the CIELab color space, the color contrast is generally expressed formula (7)

$$\left|\delta_{x,y}\right| = \sqrt{\left(L_x - L_y\right)^2 + \left(a_x + a_y\right)^2 + \left(b_x + b_y\right)^2} \tag{7}$$

After the deduction by Lu et al. [28], the energy function (7) can be rewritten as formula (8)

$$E(g) = -\sum \ln\left(N_\sigma\left(\Delta g_{x,y} + \delta_{x,y}\right) + N_\sigma\left(\Delta g_{x,y} + \delta_{x,y}\right)\right) \tag{8}$$

In formula (8), $\Delta g_{x,y} = g_x - g_y$, denotes the gray difference between the point A and the point B. When the linear combination parameters (w_r, w_g, w_b) of gray level g make the value of $E(g)$ in formula (8) minimum, the contrast information in the original image is retained to the greatest extent.

There are infinite combinations of (w_r, w_g, w_b) combinations satisfying the restraint condition. Considering that the small change of coefficient has little influence on the result, the parameter only need to take the value of 0.1 interval between [0, 1]. Therefore, (w_r, w_g, w_b) only have 66 kind of values. Through exhaustive method, the linear decoloring parameters (l_R, l_G, l_B) with the most contrast information can be found out from 66 candidates.

In the process of calculating (l_R, l_G, l_B), in order to reduce the time consumption, high-resolution input image can be reduced to low-resolution image, and get the corresponding parameter values in low-resolution images.

The grayscale image obtained by using the method of contrast preserving is shown in Fig. 5(c).

By observing three images in Fig. 5, compared with the Fig. 5(b), Fig. 5(c) can reflect the difference between different colors in the original image (Fig. 5(a)) more accurately.

From the perspective of visual effects, the result of grayscale processing method of preserving the contrast are more consistent with the visual effect of human.

Combining formula (3) and formula (8), contrast value L based on the difference of RGB channels among regions can be expressed by formula (9)

$$L = D_R * l_R + D_G * l_G + D_B * l_B \tag{9}$$

4 Experimental Results

In order to verify the performance of our method, we counted the time spent in experiments. The computer we used to do the experiments is model MSI micro star GP62 2QE-215XCN, with the configuration of Intel Core i7 5700HQ CPU, 2.7 GHz main frequency, 8G memory, and dual graphics card, the size of experimental image is 640 * 480. By statistic, the average time of image processing is about 92 ms.

In order to verify the rationality of our method, the calculation result of our contrast method (L) of the experimental image in Fig. 3 is shown in Fig. 6.

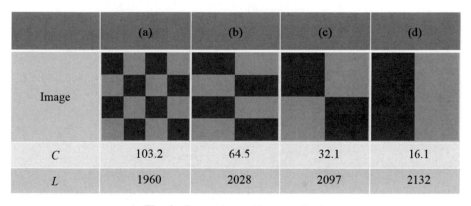

Fig. 6. Comparison with our method.

From Fig. 6. The GLCM method have a large numerical change rate, this is unreasonable; and our method have a small numerical change rate, this is more consistent with the objective fact that the experimental images are similar.

In order to verify the advantage of our method, we calculate the contrast value L of the natural environment image in Fig. 1, as shown in Fig. 7, and analyzed in the aspect of subjective vision.

In Fig. 8, from the subjective visual effect, the yellow object on the left image is obviously different from the background, and the texture details are clear. The visual effect of the image is better than that of the right image with obvious blurred phenomenon. But the GLCM value of the left image is lower than that of the right image, this is inconsistent with the subjective visual observation results. The contrast value calculated by our method shows that the contrast value in the left image is much greater than that of the right image, this is consistent with the visual observation results.

(a) C=61.5; L=179.2   (b) C=88.7; L=88.1

Fig. 7. Results of different contrast.

In order to verify the effectiveness of our method, the proposed method was compared with multiple methods, i.e., The methods used include: GLCM [21], RMS [25], SD [22], AQOC [8]. We select some natural environment images and calculate the contrast value. The results are shown in Fig. 8.

Image	GLCM	RMS	SD	AQOC	Our method	Image	GLCM	RMS	SD	AQOC	Our method
	13.8	14.5	1830	0.15	323.9		63.8	68.0	3550	0.20	411.3
	88.7	24.6	761	0.09	88.1		20.7	20.7	969	0.11	95.3
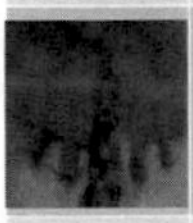	6.0	11.4	604	0.07	76.3		21.0	9.5	1141	0.11	108.5
	223.9	77.7	5615	0.26	626.2		71.1	59.8	3812	0.22	453.8
	1.47	3.6	346	0.06	42.2		3.4	4.7	470	0.07	53.3

Fig. 8. The contrast calculation results of different methods for different images. (Color figure online)

In Fig. 8, we show the calculation results of different contrast methods. From the calculated contrast value, GLCM method and RMS method have individual data items that are quite different from other methods, this is because they are affected by uniform or nearly uniform color blocks in the original image; From the contrast value of our method and other methods, the value trend of the image contrast value is similar. When there are obvious bright areas in the image (such as images in the fourth row in Fig. 8), the corresponding contrast values of each method will be higher. When there is obvious color bias in the image (such as images in the fifth row in Fig. 8), the corresponding contrast values of each method will be lower. This shows that our method can effectively reflect the information of image contrast.

5 Conclusion

Aiming at the calculation error of the traditional global contrast method for the original image with a large uniform or nearly uniform color blocks. This paper presents a contrast calculation method based on RGB-difference among regions. Firstly, divide the image into several regions to obtain the difference information of three channels in different regions of the image. At the same time, we perform grayscale processing on the image, and obtain the proportion value of the grayscale processing parameters which save the contrast information of the original image. Finally, combine the difference information of the three channels among regions with the grayscale processing parameters to get the contrast information of the original image. The verification results show that compared with the widely used traditional global contrast method, this method has less affected by the uniform or nearly uniform color blocks in original image, and can reflect the image contrast information more objectively and fairly. In addition, the contrast information obtained by this method is consistent with the observation results of human vision on the image quality, which conforms to the objective law.

However, when some regions of the original image are severely exposed, the contrast value of this method will be higher, this is unreasonable. In response to this phenomenon, we still need to do more optimization research in future work.

References

1. Kajla, V., Gupta, A., Khatak, A.: Analysis of X-ray images with image processing techniques: a review. In: International Conference on Computing Communication and Automation, pp. 1–4 (2018)
2. Dutta, A., Dubey, A.: Detection of liver cancer using image processing techniques. In: International Conference on Communication and Signal Processing, pp. 315–318 (2019)
3. Yang, M., Sowmya, A.: An underwater color image quality evaluation metric. IEEE Trans. Image Process. 24, 6062–6071 (2015)
4. Mohamed, N.M.A., Lin, L.Q., Chen, W.L., Wei, H.A.: Underwater image quality: enhancement and evaluation. In: Cross Strait Radio Science & Wireless Technology Conference, pp. 1–3 (2020)
5. Ahmed, I.T., Der, C.S., Jamil, N., Hammad, B.T.: Analysis of probability density functions in existing no-reference image quality assessment algorithm for contrast-distorted images. In: Control and System Graduate Research Colloquium, pp. 133–137 (2019)
6. Wardani, K., Kurniawan, A., Mulyana, E., Hendrawan: ROI based post image quality assessment technique on multiple localized filtering method on kinect sensor. In: International Conference on Telecommunication Systems, Services, and Applications, pp. 1–4 (2018)
7. Sethakamnerd, P., Leevailoj, C.: Comparison of contrast ratio of two ceramics in two different thicknesses. In: International Conference on Science and Technology, pp. 105–109 (2015)

8. Yelmanov, S., Romanyshyn, Y.: A quick no-reference quantification of the overall contrast of an image. In: 2020 IEEE Third International Conference on Data Stream Mining & Processing, pp. 185–190 (2020)
9. Kim, Y., Koh, Y.J., Lee, C., Kim, S., Kim, C.S.: Dark image enhancement based on pairwise target contrast and multi-scale detail boosting. In: International Conference on Image Processing, pp. 1404–1408 (2015)
10. Beghdadi, A., Qureshi, M.A., Amirshahi, S.A., Chetouani, A., Pedersen, M.: A critical analysis on perceptual contrast and its use in visual information analysis and processing. IEEE Access **8**, 156929–156953 (2020)
11. Kim, J.U., Ro, Y.M.: Attentive layer separation for object classification and object localization in object detection. In: International Conference on Image Processing, pp. 3995–3999 (2019)
12. Barten, P.G.: Contrast Sensitivity of the Human Eye and Its Effects on Image Quality. SPIE, Bellingham (1999)
13. Kanimozhi, S., Gayathri, G., Mala, T.: Multiple object identification using single shot multibox detection. In: International Conference on Computational Intelligence in Data Science, pp. 1–5 (2019)
14. Sai, B.N.K., Sasikala, T.: Object detection and count of objects in image using tensor flow object detection API. In: International Conference on Smart Systems and Inventive Technology, pp. 542–546 (2019)
15. Agaian, S.S., Silver, B., Panetta, K.A.: Transform coefficient histogram-based image enhancement algorithms using contrast entropy. IEEE Trans. Image Process. **16**, 741–758 (2007)
16. Triantaphillidou, S., Jarvis, J., Psarrou, A., Gupta, G.: Contrast sensitivity in images of natural scenes. Signal Process. Image Commun. **75**, 64–75 (2019)
17. Lillesaeter, O.: Complex contrast, a definition for structured targets and backgrounds. J. Opt. Soc. Am. A **10**, 2453–2457 (1993)
18. Cornsweet, T.N.: Visual Perception. Academic (1970)
19. Fechner, G.T.: Elemente der psychophysik (leipzig: Breitkopf & hartel). In: English Translation, vol. 1. Winston, New York (1860)
20. Calabria, A.J., Fairchild, M.D.: Perceived image contrast and observer preference II. Empirical modeling of perceived image contrast and observer preference data. J. Imaging Sci. Technol. **47**, 494–508 (2003)
21. Haralick, R.M., Shanmugam, K., Dinstein, I.: Textural features for image classification. IEEE Trans. Syst. Man Cybern. 610–621 (1973)
22. Moulden, B., Kingdom, F., Gatley, L.F.: The standard deviation of luminance as a metric for contrast in random-dot images. Perception **19**, 79–101 (1990)
23. Rizzi, A., Algeri, T., Medeghini, G., Marini, D.: A proposal for contrast measure in digital images. In: Conference on Colour in Graphics, pp. 187–192 (2004)
24. Shaus, A., Faigenbaum-Golovin, S., Sober, B., Turkel, E.: Potential contrast—a new image quality measure. In: Proceedings of the IS&T International Symposium on Electronic Imaging 2017, pp. 52–58 (2017)
25. Bex, P.J., Makous, W.: Spatial frequency, phase, and the contrast of natural images. J. Opt. Soc. Am. A **19**, 1096–1106 (2002)
26. Calculate image contrast. https://blog.csdn.net/weixin_45342712/article/details/96591834. Accessed 13 Apr 2021
27. Wang, Y.F., Huang, Q., Hu, J.: Adaptive enhancement for low-contrast color images via histogram modification and saturation adjustment. In: International Conference on Image, Vision and Computing, pp. 405–409 (2018)
28. Yelmanov, S., Romanyshyn, Y.M.: Automatic contrast enhancement of complex low-contrast images. In: International Conference on Advanced Trends in Radioelecrtronics, Telecommunications and Computer Engineering, pp. 952–957 (2018)

Virtual Reality and Human-Computer Interaction

Research on Key Technologies and Function Analysis of Live Interactive Classroom in AI+ Era

Zhou Nan[1,2,3](✉) and Zhou Jianshe[3]

[1] School of Literature, Capital Normal University, Beijing 100048, China
zhoun@bjou.edu.cn
[2] Intelligence Education Institute, Beijing Open University, Beijing 100081, China
[3] China Language Intelligence Research Center, Capital Normal University, Beijing 100048, China

Abstract. Intelligent live interactive classroom in the era of AI+ is a quality assessment method of precision teaching based on deep learning and big data behavior analysis. It has gradually developed into an important teaching approach and method. On the issues of teaching standard missing in adult open education online live class, education quality supervision is not in place and lack of interactivity, this paper explores the key technologies of live interactive teaching platform for open education, and puts forward the design method of AI+ live classroom framework and functions from the aspects of teaching management and evaluation, so as to realize learner-centered comprehensive intelligence. This method plays an important role in promoting the individualized teaching and accurate service of open education.

Keywords: The age of AI plus · Open education · Simultaneous live classroom · Intelligent teaching

1 Introduction

Broadcast online classroom refers to the teaching activities that can be carried out anytime and anywhere through computers or mobile devices. It is a new generation teaching mode of "Internet plus Traditional Education" generated with the rapid development of internet technology [1–3]. According to statistics, in 2016, the number of online education users in China was 90.14 million, with a market size of 150.62 billion yuan. It is expected to reach 160 million in 2019, with a market size of more than 260 billion yuan. Compared with traditional school learning and online learning, live education provides a more open platform for students and teachers, which helps break through time and space constraints, enhances the sharing of educational resources, promotes educational equity, and solves many problems existing in traditional offline education. This study mainly focuses on the changes AI has made to the online live-streaming teaching platform under the background of adult higher education, elaborates the problems existing in the current adult higher education industry, analyzes the present situation of live online teaching

© Springer Nature Singapore Pte Ltd. 2021
Y. Wang and W. Song (Eds.): IGTA 2021, CCIS 1480, pp. 203–213, 2021.
https://doi.org/10.1007/978-981-16-7189-0_16

platform and mainly introduces the application and role of AI technology in the synchronous live broadcast platform, and provides feasible suggestions for the development direction of "AI+ live broadcast teaching" in the future.

2   Advantages and Disadvantages of Webcast Classroom

2.1   Advantages of Webcast Class

1) **Break the time and space limit of education**

The combination of traditional education and Internet technology has also completely changed the traditional offline education model, bringing revolutionary changes to traditional education. The breakthrough of time and space in synchronous live class is reflected in two aspects: time and space. In terms of time, simultaneous broadcast classroom teaching does not require students to go to offline teaching classes, which greatly saves students' time cost. In terms of space, live teaching classes do not require students to go to the teaching place. Online learning can be conducted at any place with Internet access at any time. Recorded courses can also be learned after downloading, which is not limited by time and space, and can help some professional workers solve time pressure to a large extent.

2) **Promoting the sharing of quality educational resources**

A prominent feature of online broadcast teaching is to promote the sharing of high-quality teaching resources, which helps to promote the fairness of social education. However, online broadcast classroom breaks the restriction of closed teaching resources, which enables all the people with Internet access to have the opportunity to enjoy high-quality teaching, and enables famous teachers to spread better teaching results to a wider range of people without considering the limitations of the school site.

3) **Reduce the cost of education**

There are also many problems in online live classroom education, among which the two most important ones are the supervision of teaching quality and the interaction between teachers and students.

2.2   Disadvantages of Webcast Class

1) **Education quality supervision is not in place**

Both offline and online education, classroom teaching and examination evaluation, education quality supervision problem widely exists in every link of adult higher education, the main problem is the student's learning environment variety, students' learning attitude and learning quality is difficult to guarantee, the learning process is also difficult to be effectively monitored.

At the same time, current education and teaching evaluation pays more attention to process assessment, and students' daily performance in class is often taken as an important indicator of assessment. It is basically impossible for live class to monitor the learning process of students, and it can only be carried out by examination. Therefore, the evaluation results may be accidental or not objective.

2) **The interaction is relatively poor**

In the process of teaching, the interaction between teachers and students is of great significance for both sides to grasp the teaching and learning progress and understand the learning situation of students. Face-to-face classroom teaching can strengthen the communication between teachers and students, solve students' problems in real time, fully guarantee teachers' teaching progress and monitor students' learning quality.

However, the synchronous teaching platform is mainly dominated by teachers, who can teach courses and answer questions to students in various forms such as video, voice and text, while students can only interact with each other through text, thus the communication efficiency between teachers and students is relatively low. In addition, the number of students is large, the quality of the questions is difficult to guarantee, and some simple questions occupy the whole discussion area, which consumes a lot of teachers' time. These situations greatly affect the interaction between teachers and students in teaching and reduce the efficiency of communication between teachers and students.

3   AI Plus Key Technology of Synchronous Live Broadcast in Class

The integration of Internet technology and offline classroom teaching enriches the forms of education and teaching, accelerates the circulation of educational resources, breaks the space-time limitation of traditional teaching, and is of great significance to promoting educational equity [4]. At present, artificial intelligence technology is in a new stage of continuous development. Many new technologies are constantly applied in production and living practice. In the development of education industry, they are mainly applied to the following technologies.

3.1   Face Recognition Technology

The most direct application of face recognition technology in AI classroom is to realize automatic classroom check-in. By comparing the image information collected by the camera with the pre-stored face information in the database, the automatic face retrieval is realized. The main technical route of face recognition is transformed from manually designed features and classification recognition to end-to-end autonomous feature learning based on DCNN, as shown in Fig. 1. DCNN uses BP algorithm for supervised learning. BP algorithm is the core algorithm of deep network training. It uses the chain derivative rule to solve the weight gradient of the objective function with respect to the multi-layer neural network. DCNN is designed to process multi-dimensional data such as images. It uses four key ideas to make use of the attributes of natural signals: local connection, weight sharing, pooling and multiple network layers. Different from artificially designed features (LBP, etc.), DCNN can autonomously learn high-level and abstract feature expression vectors from end to end. For input of multidimensional face, with the increase of the depth of the neural network, convolution and pooling layer upon layer overlay, the number of neurons is the corresponding decrease, eventually form a specific, compact, low dimension, the global face feature expression vector (usually the second from bottom of hidden layer) is used for face recognition (by KNN classifier, etc.), Face verification (calculating distance) and other tasks.

Fig. 1. DCNN network structure

3.2 Human Behavior and Expression Analysis

Human behavior analysis and expression analysis are mainly used in classroom management and monitoring students' learning status. Based on 3D convolutional neural network, the characteristics of human behavior in time and space dimension were extracted from the surveillance video, and the abnormal behaviors of students were inferred and analyzed, which provided an evaluation basis for the automatic evaluation system. In addition, facial expression analysis, on the one hand, can infer students' learning attitude in class; on the other hand, facial expression analysis can solve students' understanding of problems through current analysis, so as to remind teachers to control the pace of class in real time and understand students' knowledge mastery.

There are two main research methods of behavior and expression recognition: one is based on manual feature extraction, and the other is based on deep network feature learning, of which the latter is the current research hotspot and development trend. The method of feature representation based on deep network learning is to automatically learn features from original data, this method is end-to-end, input video, output classification results. The deep networks used for behavior recognition in deep learning mainly include convolutional neural network (CNN) and recursive neural network (RNN). Convolutional neural networks usually follow a three-layer architecture, namely, the convolutional layer, the pooling layer and the fully connected layer. The classic is the two-stream CNN proposed by Simonyan et al. [5]. for behavior recognition, as shown in Fig. 2. Recursive neural network is also commonly used in deep learning models. It takes data of previous moments as data input of the current moment, so as to retain information in time dimension. LSTM (Long Short-Term Memory) type RNN is an extension of ordinary RNN. Niebles et al. [6] proposed an unsupervised LSTM model to calculate the presentation information of video.

3.3 Natural Language Processing

Natural language processing technology is mainly used in the teacher-student interaction process, students are relatively simple questions to automatically reply. Through the LSTM network +CRF statistical method, understand the text semantics, and conduct online search at the same time, match the answers to questions, help students answer

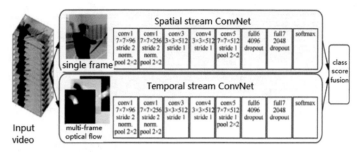

Fig. 2. The two-stream CNN workflow

simple questions, and improve the efficiency of communication between teachers and students.

In the past three years, most applications in natural language processing have used deep learning to solve problems. In 2017, Kadari et al. [7] proposed a method to solve the CCG hypertag task. By combining bidirectional long-term short-term memory and conditional random field (BLSTM-CRF) model, this method extracted input features and completed labeling, and achieved excellent results. In 2017, Kim et al. [8] proposed a neural network frame diagram that uses sets to perform dependency analysis on natural statements. The set method assigns the sliding input position to a component classifier containing the label position to be predicted. In 2017, Xiao et al. [9] proposed a Chinese sentiment classification model based on the concept of convolutional control block (CCB). Based on the CCB model, this method takes the sentence as the unit, considers the short-term and long-term contextual dependence for emotion classification, connects the word participle in the sentence into five layers of CCB, and achieves a good prediction accuracy of 92.58% for positive emotion. In 2018, Wu et al. [10] proposed a hybrid unsupervised method to solve the two important tasks of long-term extraction (ATE) and opinion objective extraction (OTE) in sentiment analysis.

3.4 Knowledge Graph and Expert Systems

The knowledge graph is widely applied. It can not only be applied to the automatic question-answering assistance to solve students' classroom problems, but also be combined with the expert system to make decisions on various things in the system, such as students' classroom performance evaluation and teachers' teaching evaluation, through other technical means such as identification and behavior analysis, as shown in Fig. 3.

In 2005, researchers from Dalian University of Technology for the first time described the basic concepts and data algorithms of knowledge graph, as well as the new progress and application prospect in the field of knowledge graph. Since then, knowledge graph has entered the field of vision of researchers in China. The commonly used knowledge graph construction methods include: bibliometrics, citation analysis, co-citation analysis, multivariate statistical analysis, word frequency analysis and social network analysis.

Chen Yulin et al. [11] analyzed the knowledge graph in the field of educational technology and concluded that He Kekang et al., through the study of author co-citation network, was the core author of educational technology. Yang Guoli et al. [12] used

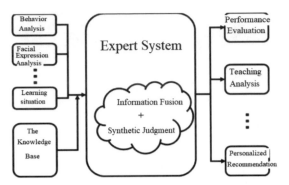

Fig. 3. Application of artificial intelligence expert system

CiteSpace software to construct the knowledge map of library science in China, explored the core institutions and authors of library science, and explained the knowledge base, frontier and hot spot of this field. Wang Youmei et al. [13] explored the research hotspots of MOOC in China through multivariate statistical analysis and put forward research suggestions. Li Changchun et al. [14] analyzed the hot current situation and development trend of network education based on knowledge graph. Ding Xueyang et al. [15] analyzed the research hotspots and future trends of educational equity in China based on the knowledge graph of coword matrix.

4   The Main Functions of Open Education AI Live Class

Compared with closed education, open education is a type of education supported by the educational concepts of "openness, inclusiveness and lifelong". It realizes the value concept of lifelong education that "everyone, everywhere and all times" can learn through the all-round openness of the objects, resources, process and management of higher education. The online live class platform of open education mainly reflects the relevant concepts of open education.

4.1   Basic Framework of the Platform

AI live interactive classroom adopts AI plus Internet live classroom to help realize remote teaching and intelligent classroom management. The platform mainly adopts mature artificial intelligence technologies such as face recognition technology, behavior recognition, natural language processing and intelligent recommendation technology, etc. The main system framework is shown in Fig. 4.

The teaching resource cloud platform can integrate and share the rich teaching resources. To realize the learning space for everyone, high-quality resources for all classes.

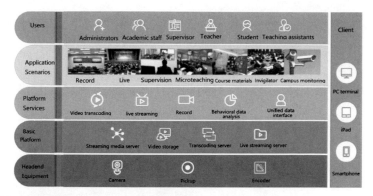

Fig. 4. Schematic diagram of AI live platform architecture of Open Education

4.2 Introduction of AI Live-Streaming Classroom Functions

Open education teaching platform of the live interaction mainly from the aspects of teaching tube evaluation to realize learner centered comprehensive intellectualized, implement teaching aspects: intelligent digital teaching resource, diversified teaching environment and interaction teaching, network of online courses, intelligent data statistics and analysis, multi-campus interactive teaching and learning aspects of implementation: Interactive learning space, online learning platform, intelligent classroom interaction, learning behavior collection, teaching resource sharing and other functions. In terms of management, centralized network management, Internet of things environment control, face recognition and identity authentication, information release, intelligent space management, artificial intelligence analysis, and in terms of evaluation: Teaching course evaluation, online course resource sharing, teacher teaching evaluation, student learning evaluation, data statistics and analysis and other functions.

1) **One place to teach, many places live, repeated replay**

"More than one to teach, to live, repeats" is inheriting the advantages of the original Internet broadcast live online, teachers and students online at the same time, teachers in the teaching, the remote broadcast system will be taught the audio and video data transfer to the stream media encoder, compressed into data flow, and synchronous transmission by streaming media server through the Internet to the receiver class, Students in many places can watch the video of different teaching at the same time online, and can interact with the live teaching. At the same time, students who fail to participate in the live lecture in time can also watch the video recorded in the live class after class to consolidate the learning of classroom knowledge, as shown in Fig. 5.

2) **Multi-channel synchronous live broadcast**

Compared with the traditional dual-channel synchronous live broadcast, there are only two information transmission channels: teacher's image and courseware playback. This platform adds a video transmission channel for teachers writing on blackboard to realize three channels of synchronous video transmission of "teacher,

Fig. 5. Teaching resources platform display

courseware and blackboard writing". Technically, three-channel multimedia transmission is similar to two-channel multimedia transmission, but it requires more network services in the process of synchronous transmission.

3) **Intelligent online discussion**

Compared with the traditional online discussion and speech process, this platform continues to make the following improvements while retaining the application functions of the original live broadcast platform:

One is to retain the original text communication mode, while expanding the audio and video questioning mode. The original text communication mode can allow teachers to give timely feedback on some valuable questions in the process of explanation, as well as in the process of questioning and communication, so as to help students to understand quickly and to avoid confusion in the process of questioning in targeted explanation. Extended audio and video modes enable students to communicate with each other through audio or video in a timely manner for some valuable and complex problems, thus improving communication efficiency and saving time and cost.

The second is to add the self-help question and answer function, aiming at some simple question platform, through BLSTM-CRF technology, analyze and identify the content of the question, and the system platform will automatically answer. If you can't answer the questions, you can communicate with the teacher via text, voice or video.

The intelligent interaction mode can greatly reduce the teacher's interaction burden and improve the interaction efficiency.

4) **Intelligent classroom management and teaching quality control**

Online teaching is the most difficult monitoring content is aimed at students in teachers' learning attitude and learning action, intelligent AI management system provides a set of feasible solutions, through the automatic monitoring system, AI the classroom DeepFace of face recognition, behavior identification technology, such as RCNN automatically evaluate the behavior of college students. Real-time video monitoring to remind students, and online evaluation.

5) **Intelligent evaluation of teaching process**

AI intelligent live broadcasting platform can not only automatically monitor the teaching process, but also objectively evaluate the teaching process through AI technology, including the evaluation of teachers and students, and assist classroom management.

AI platform can make comprehensive evaluation based on teachers' class behavior, teaching and explanation, student evaluation and other aspects. Compared with other subjective evaluations, AI intelligent evaluation system is more objective and fair.

For students of evaluation mainly includes the main performance of the students in class and record the process of computer operation, such as the existence of watching video, whether in the treatment of other files, and whether to take an active part in class discussions, etc., through the behavior recognition, emotion recognition analysis, eyeball is over, on-line monitoring students' learning attitude, learning style and application of knowledge, Give the students a grade evaluation in class, and at the same time remind the students to participate in the class seriously, as shown in Fig. 6.

6) **Personalized recommendation**

Individualized training and adaptive education is an ideal mode of talent training, and also a means and method to solve the confusion of youth in the process of accepting adult higher education. Adaptive education and personalized learning are AI teaching platforms, which use the acquired learning data of different students to provide personalized training programs for students with different learning progress and learning ability through knowledge mapping and other methods, and provide targeted guidance for students to help them improve their ability level to the maximum.

Fig. 6. Intelligent analysis of teaching process

The AI teaching platform can recommend learning courses according to students' learning data such as class behavior response, learning progress, learning ability and learning status. At the same time, personalized training can be carried out according to students' needs. Personalized education can maximize the value of adult education, and prevent students with good foundation from wasting time in simple tasks, and students with poor foundation from failing to keep up with the pace of learning. Let every student who accepts higher education learn something, learn something, let adult education play the maximum value.

5 Conclusion

This study explores the era of AI, AI technology and Internet technology under the background of the development of network broadcast classroom interaction research problem, to the current problems existing in the development process of adult education for simple summary, but the AI plus Era the application of open education is still relatively shallow, in fact the AI plus Era of open education teaching ways and teaching methods more rich. AI live class is more intelligent and more convenient for interaction.

Acknowledgment. This project supported by the following Projects: The National Natural Science Foundation of China (Nos. 61871028); Beijing Natural Science Foundation (KZ201951160050); and Funding Project for Academic Human Resources Development in Institutions of Great Wall scholars project of Beijing Municipality (CIT&TCD20190313).

References

1. Zhu, Z.T., Peng, H.C.: Deep Learning: the core pillar of intelligent education. J. Chin. Soc. Educ. (5), 36–45 (2017)
2. Liu, J.: Live + education: a new form of "Internet Plus" learning and its value inquiry. J. Distance Educ. (1), 52–59 (2017)
3. Hao, C.E.: Reform of Chinese education pattern caused by live education. Chin. J. ICT Educ. High. Educ. Vocat. Educ. (12), 28–29 (2016)
4. Fu, C.R.: Artificial intelligence, to creat personalized and customized education. J. East China Normal Univ. Educ. Sci. **35**(5), 13–14 (2017)
5. Simonyan, K., Zisserman, A.: Two-stream convolutional networks for action recognition in videos. Neural Inf. Process. Syst. **1**(4), 568–576 (2014)
6. Niebles, J.C., Wang, H., Li, F.F.: Unsupervised learning of human action categories using spatial-temporal words. Int. J. Comput. Vis. **79**(3), 299–318 (2008)
7. Kadari, R., Zhang, Y., Zhang, W., et al.: CCG Supertagging via bidirectional LSTM-CRF neural architecture. Neurocomputing **283**, 31–37 (2017)
8. Kim, K., Jin, Y., Na, S.H., et al.: Center-shared sliding ensemble of neural networks for syntax analysis of natural language. Expert Syst. Appl. **83**, 215–225 (2017)
9. Xiao, Z., Li, X., Wang, L., et al.: Using convolution control block for Chinese sentiment analysis. J. Parallel Distrib. Comput. **116**, 18–26 (2017)
10. Wu, C., Wu, F., Wu, S., et al.: A hybrid unsupervised method for aspect term and opinion target extraction. Knowl.-Based Syst. **148**, 66–73 (2018)

11. Chen, Y.L., Run, Z.M.: Construction and analysis of visualized knowledge graph of academic groups of educational technology in China. China Educ. Technol. (12), 1–7+13 (2012)
12. Yang, G.L., Li, P., Liu, J.: Analysis on knowledge graph of library science research in China. J. Natl. Libr. China (1), 52–59 (2012)
13. Wang, Y.M., Ye, A.M., Lai, W.H.: What's the future of MOOCs: analysis of domestic research hotspots based on knowledge graph. China Educ. Technol. (07), 12–18 (2015)
14. Zhu, C.C., Li, N.: Research hotspots and trends of online education resource sharing: based on knowledge graph analysis. Cult. Educ. Mater. (2017)
15. Ding, X.Y., Cheng, T.J.: Hotspots and future trends of educational equity research in China since the 21st century-analysis of knowledge graph based on co-word matrix. J. Distance Educ. **528**(01), 13–21+50+96 (2019)

Applications of Image and Graphics

BrainSeg R-CNN for Brain Tumor Segmentation

Jianxin Zhang[1,2,3(✉)], Xinchao Cheng[2], Tao He[1,3], and Dongwei Liu[1,3]

[1] School of Computer Science and Engineering, Dalian Minzu University,
Dalian, China
[2] Key Lab of Advanced Design and Intelligent Computing (Ministry of Education),
Dalian University, Dalian, China
[3] SEAC Key Laboratory of Big Data Applied Technology, Dalian Minzu University,
Dalian, China

Abstract. Brain tumor segmentation methods using deep neural networks have recently achieved significant performance breakthroughs. However, the existing brain tumor segmentation networks are directly implemented on whole brain images, resulting in possibly reduced segmentation performance due to the disturbance of background regions. To solve this problem, inspired by the Mask R-CNN, a novel brain tumor segmentation model called BrainSeg R-CNN is proposed in this work, which classifies the brain tumor areas and boundaries based on the detected region of interest in an end-to-end manner to achieve segmentation result. Also, an effective feature extraction strategy is presented in BrainSeg R-CNN, which in detail extracts various kinds of information from separate channels for each modality and immediately adopts a cross-connection operator to realize the information transmission among different channels. Moreover, concatenation and add calculation are integrated to improve the fusion efficiency of multi-scale features from brain tumor images. Additionally, a multi-weighted and multi-task loss function which fully considers tumor size and overlap label is introduced, significantly improving the segmentation performance. Experimental results on BraTS 2017 dataset demonstrate that our BrainSeg R-CNN obtains competitive performance with state-of-the-arts.

Keywords: Convolutional neural network · Segmentation · Brian tumor · BrainSeg R-CNN

1 Introduction

With the rapid development of deep learning in the field of medical imaging, brain tumor segmentation task, as a key step in brain function analysis and disease diagnosis, has also made a major breakthrough in recent years [1,2]. The initial deep segmentation networks take brain tumor segmentation as patch classification problem, mainly employing typical convolutional neural networks (CNN) architectures in visual classification task. Also, sliding window and post processing are adopted to achieve the entire segmentation result. The main

© Springer Nature Singapore Pte Ltd. 2021
Y. Wang and W. Song (Eds.): IGTA 2021, CCIS 1480, pp. 217–226, 2021.
https://doi.org/10.1007/978-981-16-7189-0_17

disadvantages of these methods lie in redundant calculations and global information loss. Then, fully convolutional networks (FCNs) are introduced to provide a pixel-to-pixel solution for brain tumor segmentation with effective expansion of receptive fields, leading to the superior segmentation accuracy and efficient calculation reduction [3,4]. Specially, an evolutionary version of FCN called UNet [5,6], which well integrates high-level and low-level features of medical images and achieves significant performance improvement in a variety of medical segmentation tasks, gradually becomes the mainstream of brain tumor segmentation methods. To further improve its segmentation performance, residual module [7], attention mechanism [8] and multi-scale fusion cascade ideology [9] are injected into the baseline model, which largely promotes the development of brain tumor segmentation methods. Although promising segmentation performance has been achieved, existing brain tumor segmentation networks [10–13] are directly performed on whole images, resulting in possibly reduced segmentation performance due to the disturbance of background regions.

To resolve this problem, inspired by the recent Mask R-CNN [15], a small and flexible object instance detection network with a segmentation branch for natural images, we propose a novel brain tumor segmentation model named BrainSeg R-CNN in this work. BrainSeg R-CNN classifies brain tumor areas and boundaries based on the detected region of interest (RoI) in an end-to-end manner to achieve segmentation result, providing a new pipeline for brain tumor segmentation. In addition, an effective feature extraction strategy is given in BrainSeg R-CNN, and it in detail extracts various kinds of information from separate channels for each modality with cross-connection operator to realize the information transmission among different channels. Also, concatenation and add calculation are integrated to improve the fusion efficiency of multi-scale features from brain tumor images. Moreover, a multi-weighted and multi-task loss function which fully considers tumor size and overlap label is introduced, and it significantly improves the segmentation performance. The proposed BrainSeg R-CNN is extensively evaluated in the brain tumor segmentation challenge (BraTS) [16], and experiment results illuminate that it gains competitive performance with state-of-the-arts. Specially, it achieves the whole tumor segmentation accuracy of 91.54% in slices with brain tumors. The overall architecture of the proposed BrainSeg R-CNN is illustrated in Fig. 1. The main contributions of this work are three folds: (1) A novel brain tumor segmentation network called BrainSeg R-CNN is proposed, which significantly distinguishes from the existing networks for this task. (2) BrainSeg R-CNN introduces effective feature extraction and fusion strategies as well as an effective loss function for brain tumor segmentation, largely improving the performance of the network. (3) Experimental results on a widely used dataset demonstrate its competitive performance with state-of-the-arts.

2   Method

The BrainSeg R-CNN is mainly inspired by the Mask R-CNN to provide a novel pipeline for brain tumor segmentation task. It adopts the similar two-stage

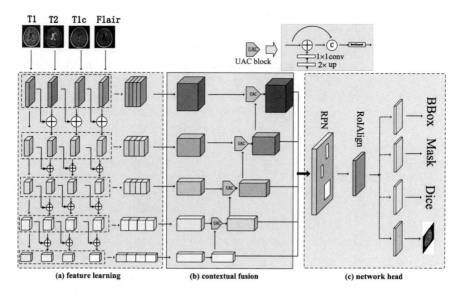

Fig. 1. Overview of BrainSeg R-CNN. It is mainly comprised by feature learning, contextual fusion and network head. It employs multi-channel and cross-modality connection to extract more discriminate features, followed by an improved feature pyramid structure for contextual fusion. An extra Dice loss is introduced on the top of network in parallel with other losses.

procedure as Mask R-CNN. Differently, as shown in Fig. 1, our BrainSeg R-CNN consists of three different parts, i.e., feature learning, contextual fusion and network head, aiming at gaining superior performance for this task.

2.1   Mask R-CNN

Here, we briefly review the Mask R-CNN [15] that is highly related to our work. Mask R-CNN takes advantage of the principle of Faster R-CNN [17] while introducing the extra mask branch so that it can predict object mask on RoI generated by region proposal network (RPN) for fast instance segmentation. Besides, Mask R-CNN improves the coarse spatial quantization of RoIPool in Faster R-CNN and alternatively proposes the quantization-free layer RoIAlign for avoiding misalignment. Mask R-CNN has provided strong baselines for multiple vision tasks such as human poses estimation and instance segmentation. As such, we follow the similar principle to deal with brain tumor segmentation task. Unfortunately, compared to natural image tasks, medical image tasks face almost very different situations, such as multi-modality images, fewer labeled samples as well as various instance shapes. Therefore, Mask R-CNN cannot be directly transferred to the brain tumor segmentation task, and we have to redesign the architecture to fit for this task.

2.2 BrainSeg R-CNN

Multi-path and Cross-Modality Feature Learning. Although four modalities (T1, T1c, T2 and Flair) contain spatially and semantically similar information, they describe brain tumor from different views and provides complementary information to each other. Effective feature learning will provide better representation of brain tumor image for following segmentation of RoI. Meanwhile, in the family of mainstream CNN models, different convolutional layers capture different visual features and varying scales information. The backbone models encode the entire input or larger feature maps spatially in lower layers, thereby harvesting finer spatial information for pixel-wise segmentation. However, due to the local convolution with small receptive fields, lower layers have poor semantic capturing capability. In higher layers, the stacked multiple convolutional layers progressively sense the entire input with larger receptive view and possess strong semantic information, but the outputs of higher layers are spatially coarse after the downsampling. Overall, the lower layers provide more accurate spatial characteristics while the high ones predict more accurate semantic labels. To this end, we design the effective features learning strategy from multi-path and cross modality, combining the inherent merits of varying convolutional layers and complementary information of four modalities.

To achieve that goal, the four modalities are separately fed into four CNN models, shown in Fig. 1(a), from left to right are T1, T1c, T2 and Flair, respectively. Motivated by the shortcut in ResNet, the features in the i-th level from T1 are combined with features in j-th ($j = i+1$) level from T2 though element-wise addition. Note that the two feature maps always have different spatial size. We conduct extra convolution with downsampling on the larger one, making them have same size. The resulting features then pass though the next convolutional layer. For other modalities, we repeat the similar operation. In this way, each modality integrates features of every level from one or more adjacent modalities except the first T1. The network not only learns features from individual CNN model and modality, but also gets multi-scale and cross-modality features, fully considering the interaction among modalities to obtain discriminative features of brain tumor. Besides, all features of the i-th level of every modality are concatenated along the channel dimension to form a new feature map to characterize brain tumor at i-th level, fed into next contextual fusion part.

Feature Pyramid Structure Based Contextual Fusion. To get better global contextual information, we present an improved feature pyramid structure to fuse features gained from feature learning period under different pyramid resolutions, depicted in Fig. 1(b). After feature learning, we get concatenated feature maps of each layer. Here the number of channels and spatial size per concatenated feature maps are different. The feature maps at deeper layers get more small spatial size with more channel number. We first perform bottleneck block on them to give them the same dimension. The UAC block is then carried out to fuse features, which primarily involves Upsampling, Add and Concatenation operations (UAC) as shown in Fig. 2.

In UAC block, given two inputted feature maps from adjacent i-th and j-th levels, denoted respectively as $\mathbf{A}$ and $\mathbf{B}$, the low resolution feature map $\mathbf{B}$ is 2× bilinear upsampled, producing feature map $\mathbf{B}^*$, to match the spatial size with high resolution $\mathbf{A}$ followed by 1×1 convolutional layer. The resulting $\mathbf{B}^*$ and $\mathbf{A}$ are added in element-wise manner, obtaining feature map $\mathbf{C} = \mathbf{B}^* + \mathbf{A}$. The added feature map $\mathbf{C}$ then are concatenated with feature map $\mathbf{A}$, getting new map $\mathbf{D} = \begin{bmatrix} \mathbf{A}, \mathbf{C} \end{bmatrix}$, which contains global and local information with stronger semantic and finer spatial resolution, particularly helpful for segmentation. Subsequently, the fused feature maps $\mathbf{D}$ are connected to one bottleneck block for feature adaption. From the deepest layer to the shallowest layer, we keep repeating above operation progressively. The outputs of all UAC blocks hold the same dimension but have different resolutions. We upsample all of them up to the same resolution as the largest with different times ratio except the shallowest one. After that, we combine them with concatenation along the channel direction. The final fused features go though vanilla RPN to generate RoI of brain tumor, and produced each RoI is fed into the network head for bounding-box recognition and mask prediction.

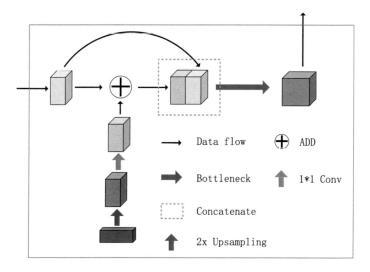

UAC block

Fig. 2. The basic structure of the given UAC block. The UAC block is designed to fuse multi-channel features, which primarily involves Upsampling, Add and Concatena-tion operations. The outputs of UAC block holds the same dimension but has different res-olutions. Additionally, it contains global and local information with stronger semantic and finer spatial resolution, which will be helpful for brain tumor segmentation.

Network Head. Our network head is similar in structure to Mask R-CNN, focusing on the guidance of training by loss function. However, due to the high similarity between tumors and tissues, their various shapes and small size, the loss function employed in Mask R-CNN actually pays too less attention on desired tumor regions, possibly resulting in poor segmentation performance and unsuitable for brain tumor segmentation task. Therefore, following [14], Brain-Seg R-CNN adds a multi-weighted loss function in conjunction with ones of Mask R-CNN in parallel fashion for brain tumor segmentation (Fig. 1(c)). The total loss is defined as following:

$$L = L_{rpn} + \lambda_1 \cdot L_{cls} + \lambda_2 \cdot L_{mask} + \lambda_3 \cdot L_{box} + \lambda_4 \cdot L_{dice} \tag{1}$$

where L_{rpn}, L_{cls} and L_{box} are identical as Mask R-CNN, which are used to train the branch of detection. L_{mask} means the average binary cross-entropy loss, and L_{dice} is the added Dice loss to optimize segmentation branch. λ_i $(1 \leq i \leq 4)$ is the hyper-parameter that controls the importance of each loss.

3   Experiments

3.1   Dataset and Settings

We evaluate the proposed BrainSeg R-CNN on the commonly used BraTS 2017 dataset. For each MRI image, there are four modalities: FLAIR, T1-weighted (T1), T1 with gadolinium enhancing contrast (T1c), and T2-weighted (T2). The dimensions of all images are $240 \times 240 \times 155$ voxels. The BraTS 2017 training set is composed of 210 cases of high-grade gliomas (HGG) and 75 cases of low-grade gliomas (LGG). Each ground-truth for brain tumors is given by experts [18,19]. Here, we divided the original training set into three subsets for model training, validation and testing, respectively. Figure 3 demonstrates two typical multi-mode brain tumor image samples in BraTS 2017 dataset.

Our experiments mainly consist of two parts: (1) Compared experiments using slices with tumors; (2) Compared experiments using all slices (whole brain image). As the BrainSeg R-CNN is based on the detection model, which will result in a higher level of false positive for slices without brain tumors. Therefore, the first part of our experiments is carried out on slices which definitely contain brain tumors to verify the effectiveness of BrainSeg R-CNN, specially to evaluate the three designed parts, i.e., feature learning, contextual fusion and network head. In the second experiment, we compare BrainSeg R-CNN with several state-of-the-art methods by using whole brain image with the same protocol as [20]. Moreover, Dice score is adopted in all of the experiments.

3.2   Compared Experiments Using Slices with Tumors

Comparison with Mask R-CNN. Here, we take Mask R-CNN architecture without multi-path and cross-modality feature (MCF), multi-scale fusion (MF) and multi-weighted dice (MD) loss as our naive baseline. Based on the different

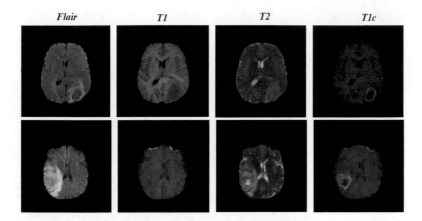

Fig. 3. Typical multi-model brain tumor images in BraTS 2017 dataset

Table 1. Comparison with Mask R-CNN [15] using slices with tumors (%)

Methods	Whole	Core	Enhance
Baseline	85.94	80.12	78.20
Baseline+MCF	86.42	82.07	80.13
Baseline+MD	87.24	82.14	80.28
Baseline+MCF+MD	87.63	82.35	80.52
Baseline+MF	88.43	83.08	80.75
Baseline+MF+MD	90.02	84.19	80.96
Baseline+MCF+MF	89.04	83.81	80.88
BrainSeg R-CNN	**91.54**	**86.22**	**81.05**

combinations of adding MCF, MF and MD, we conduct a series of comparative experiments on BraTS 2017 dataset whose results are reported in Table 1. As shown in Table 1, by introducing MCF and MF as well as the MD loss, our Brain-Seg R-CNN achieves the optimal segmentation performance of 91.54%, 86.22% and 81.05% on whole, core and enhance tumors, which outperforms that of Mask R-CNN over 5.58%, 6.10% and 2.85%, respectively. In addition, following conclusions can be drawn from Table 1. All of the MCF, MF and MD gain performance improvement over the baseline. Among them, MF is superior to the others while MCF achieves the smallest effect. Further performance improvements can be achieved through the combination of MCF, MF and MD.

Comparison with U-Net Models. Here, we mainly compare BrainSeg R-CNN with several typical 2D U-Net models including basic U-Net, Res-UNet and Res-UNet with weighted-Dice (Res-UNet+WD) on BraTS 2017 dataset to give a further evaluation, and the compared results are shown in Table 2. Table 2 illuminates that BrainSeg R-CNN achieves promising performance improvement over

Table 2. Comparison with U-Net models using slices with tumors (%). * indicates the result of our recurrence.

Methods	Whole	Core	Enhance
U-Net* [5]	83.41	80.38	76.08
Res-UNet*	86.32	84.96	79.27
Res-UNet+Weighted-Dice [7]	88.51	87.96	80.17
BrainSeg R-CNN(ours)	**91.54**	**86.22**	**81.05**

basic U-Net and Res-UNet. Compared with Res-UNet+WD, BrainSeg RCNN respectively gains 3.03% and 0.88% performance improvement on whole and enhance tumor segmentation results. Meanwhile, it is inferior to Res-UNet+WD on core tumor segmentation. However, the overall experimental results demonstrate the effectiveness of our BrainSeg R-CNN method for brain tumor segmentation.

3.3 Compared Experiments Using Whole Brain Image

To further test BrainSeg R-CNN, we compare it with several state-of-the-art methods on all slices (whole brain image) with the same setting as [20], and experiment results on BraTS 2017 dataset are given in Table 3. Among them, dense FCN (DFCN) employs typical 2D FCN model and introduces dense connection to improve the segmentation accuracy [20]. In contrast, FCN+CRF adopts 2D FCN model followed by conditional random field (CRF) as post processing [12,13]. As BrainSeg R-CNN is based on detection model, it will result in a high level of false positive for slices without brain tumors. However, this problem can be resolved by adding a pre-classifier before feature learning. Here, we take U-Net as the pre-classifier and denote this method as BrainSeg R-CNN+Classifier.

Table 3. Comparison with state-of-the-art methods using whole brain image (%)

Methods	Whole	Core	Enhance
DFCN [20]	84.00	83.00	80.00
FCN+CRF [13]	87.20	83.00	76.00
U-Net [5]	83.00	80.00	75.00
Res-UNet+WD [7]	88.13	87.36	80.12
BrainSeg R-CNN(ours)	86.54	84.88	78.49
BrainSeg R-CNN+Classifier (ours)	**91.22**	**85.62**	**80.71**

Table 3 illustrates that BrainSeg R-CNN overall outperforms DFCNN, FCN+ CRF and U-Net methods. Due to the high false positive on slices without tumors, it is inferior to the Res-UNet+WD method. However, with a simple pre-classifier as supplement, our BrainSeg R-CNN+Classifier obtains the optimal performance for both of whole and enhance tumor segmentation. Specially, it gains 91.22% Dice score for whole tumor segmentation, which is significantly higher than the others.

4 Conclusion

In this paper, inspired by Mask R-CNN, we propose a novel brain segmentation method called BrainSeg R-CNN, which classifies brain tumor areas and boundaries based on the detected RoI to finish segmentation, avoiding invalid segmentation calculation in the background area as well as providing a new pipeline for this task. Additionally, three improvements are presented in BrainSeg R-CNN to achieve better segmentation performance. Extensive experiment results on widely used brain tumor segmentation dataset demonstrate the effectiveness of our proposed BrainSeg R-CNN method. In the future, the more powerful pre-classifier will be integrated into current BrainSeg R-CNN model to further improve its performance on the entire brain image. In addition, we will extend the proposed BrainSeg R-CNN method into 3D model, and this could further avoid the wrong detection existing in 2D method.

Acknowledgements. This work was partially supported by the National Natural Science Foundation of China (61972062), the National Key R&D Program of China (2018YFC0910506), the Key R&D Program of Liaoning Province (2019 JH2/10100030), the Young and Middle-aged Talents Program of the National Civil Affairs Commission, the Liaoning BaiQianWan Talents Program, and the University-Industry Collaborative Education Program (201902029013).

References

1. Bakas, S., Reyes, M., Jakab, A., et al.: Identifying the best machine learning algorithms for brain tumor segmentation, progression assessment, and overall survival prediction in the BRATS Challenge. arXiv preprint arXiv:1811.02629 (2018)
2. Tiwari, A., Srivastava, S., Pant, M.: Brain tumor segmentation and classification from magnetic resonance images: review of selected methods from 2014 to 2019. Pattern Recogn. Lett. **131**, 244–260 (2020)
3. Shen, H., Wang, R., Zhang, J., et al.: Multi-task fully convolutional network for brain tumour segmentation. In: Annual Conference on Medical Image Understanding and Analysis (MIUA), pp. 239–248 (2017)
4. Long, J., Shelhamer, E., Darrell, T.: Fully convolutional networks for semantic segmentation. In: Proceedings of the IEEE conference on Computer Vision and Pattern Recognition (CVPR), pp. 3431–3440 (2015)
5. Dong, H., Yang, G., Liu, F., et al.: Automatic brain tumor detection and segmentation using U-Net based fully convolutional networks. In: Annual Conference on Medical Image Understanding and Analysis (MIUA), 506–517 (2017)

6. Ronneberger, O., Fischer, P., Brox, T.: U-net: convolutional networks for biomedical image segmentation. In: International Conference on Medical Image Computing and Computer-Assisted Intervention (MICCAI), pp. 234–241 (2015)
7. Kermi, A., Mahmoudi, I., Khadir, M.T.: Deep convolutional neural networks using U-Net for automatic brain tumor segmentation in multimodal MRI volumes. In: International MICCAI Brainlesion Workshop (BrainLes), pp. 37–48 (2018)
8. Zhang, J.X., Jiang, Z.K., Dong, J., et al.: Attention gate ResU-Net for automatic MRI brain tumor segmentation. IEEE Access **8**, 58533–58545 (2020)
9. Zhou, C., Ding, C., Lu, Z., et al.: One-pass multi-task convolutional neural networks for efficient brain tumor segmentation. In: International Conference on Medical Image Computing and Computer-Assisted Intervention (MICCAI), pp. 637–645 (2018)
10. Mlynarski, P., Delingette, H., Criminisi, A., et al.: 3D convolutional neural networks for tumor segmentation using long-range 2d context. Comput. Med. Imaging Graph. **73**, 60–72 (2019)
11. Tseng, K.L., Lin, Y.L., Hsu, W., et al.: Joint sequence learning and cross-modality convolution for 3d biomedical segmentation. In: IEEE Conference on Computer Vision and Pattern Recognition (CVPR), pp. 6393–6400 (2017)
12. Kamnitsas, K., Ledig, C., Newcombe, V.F.J., et al.: Efficient multi-scale 3D CNN with fully connected CRF for accurate brain lesion segmentation. Med. Image Anal. **36**, 61–78 (2017)
13. Zhao, X., Wu, Y., Song, G., et al.: A deep learning model integrating FCNNs and CRFs for brain tumor segmentation. Med. Image Anal. **43**, 98–111 (2018)
14. Sudre, C.H., Li, W., Vercauteren, T., et al.: Generalised dice overlap as a deep learning loss function for highly unbalanced segmentations. In: International Workshop on Deep Learning in Medical Image Analysis & International Workshop on Multimodal Learning for Clinical Decision Support (DLMIA & ML-CDS), pp. 240–248 (2017)
15. He, K.M., Gkioxari, G., Dollár, P., et al.: Mask R-CNN. In: IEEE International Conference on Computer Vision (ICCV), pp. 2961–2969 (2017)
16. Menze, B.H., Jakab, A., Bauer, S., et al.: The multimodal brain tumor image segmentation benchmark (BRATS). IEEE Trans. Med. Imaging **34**(10), 1993–2024 (2015)
17. Ren, S.Q., He, K.M., Girshick, R., et al.: Faster R-CNN: towards real-time object detection with region proposal networks. In: Advances in Neural Information Processing Systems (NIPS), pp. 91–99 (2015)
18. Bakas, S., Akbari, H., Sotiras, A., et al.: Advancing the cancer genome Atlas glioma MRI collections with expert segmentation labels and radiomic features. Scientific Data **4**(1), 1–13 (2017)
19. Bakas, S. Akbari, H., Sotiras, A., et al.: Segmentation labels and radiomic features for the pre-operative scans of the TCGA-GBM collection. Cancer Imaging Arch. **286** (2017). https://doi.org/10.7937/K9/TCIA.2017.KLXWJ
20. Shaikh, M., Anand, G., Acharya, G., et al.: Brain tumor segmentation using dense fully convolutional neural network. In: International MICCAI Brainlesion Workshop (BrainLes), pp. 309–319 (2017)

A Real-Time Tracking Method for Satellite Video Based on Long-Term Tracking Framework

Yufei Ding[⊠], Hongyan He, Shixiang Cao, and Yu Wang

Beijing Institute of Space Mechanics and Electricity, Beijing 100094, China

Abstract. Tracking accuracy and frame rate are the two most important indexes of satellite video tracking. Now, the research of close-range tracking algorithm is gradually developing from short-term to long-term, and has been widely used. This paper refers to the idea of long-term tracking strategy and introduces it into satellite video target tracking. With a variety of features as precision guarantee, the redetection function is added to the tracker to ensure the robust tracking. At the same time, the feature dimension is properly reduced. The validity of the algorithm is verified by the video sequence obtained from Skybox-1 satellite. The results show that the tracking effect is good, and the performance of the dimension-reduced version has no significant decrease, and the frame rate is improved by 37.8%. It shows the high feasibility and wide application prospect of long-term tracking strategy in space-based video tracking.

Keywords: Remote sensing · Satellite videos · Object tracking · Correlation filter

1 Introduction

As a new observation development direction, high resolution dynamic imaging has a wide application prospect in civil traffic monitoring, emergency response and major engineering progress [1]. With the development of space-borne imaging technology, quite a few countries and some international agencies have launched space video satellites. For instance, Skybox-1 [2], UrtheCast, and Jilin-1 all guarantee imaging quality at least sub meter level. Therefore, with the development of high spatial resolution video satellites and the increasing demand for commercial remote sensing applications, how to make full use of space-based video data is an important research direction at present and in the future.

Target tracking is the essential part of remote sensing applications of commercial video satellite, but the state-of-the-art motion tracking methods lack the adaptability to space-based imaging disturbances, which turns out to be the main constraint on the development of video field [3]. Specifically, this is mainly due to the fact that the majority of small targets in the satellite video are almost composed of a small number of pixels with similar brightness. There is almost no texture and features to be extracted inside, and it is difficult to describe and relate features. In addition, the size of satellite video image

© Springer Nature Singapore Pte Ltd. 2021
Y. Wang and W. Song (Eds.): IGTA 2021, CCIS 1480, pp. 227–237, 2021.
https://doi.org/10.1007/978-981-16-7189-0_18

is about 100 times that of ordinary video image, which puts forward higher real-time requirements for tracking algorithm. In order to solve these problems, it is necessary to study and improve the target tracking algorithm.

Correlation filtering tracking algorithm is one of the most commonly used tracking methods. It is popular among researchers because of its fast real-time performance and high accuracy. Based on this, researchers also proposed many space-based video tracking methods [4, 5], and achieved good results. However, the correlation filtering methods have problems such as negative samples leading to boundary effect, which may lead to tracking failure. Satellite video data exacerbate this problem.

Aiming at the above problems, this paper improves the correlation filtering tracking method, and proposes a long time target real-time tracking method based on space-based video. This method introduces a long-term tracking strategy which consists of tracking, learning and detection, and enables the algorithm to activate the redetection function after losing the target. Meanwhile, in order to ensure the accuracy of the algorithm, the multi-feature fusion methods of Histogram of oriented gradient (HOG) [6], Color Name (CN) [7] and Histogram of Local Intensities (HOI) are used to obtain enhanced descriptors. Finally, principal component analysis (PCA) is used to reduce the dimension of the tracker features to reduce the computational complexity.

2   Method Framework

In this paper, according to the long-term tracking mechanism, the algorithm is divided into four modules: translation filter, scale filter, long-term filter and re-detection mechanism. The overall framework of the algorithm is as follows (Fig. 1):

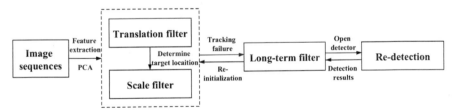

Fig. 1. The long-term tracking framework used in this paper

The details of the algorithm will be discussed in Sect. 2.1–2.3 of this paper.

2.1   Feature Extraction

The selection of features is the most important factor affecting accuracy. Compared with close-range video, the background of satellite video image is complex and changeable, so the single feature cannot guarantee the industrial requirements. Therefore, more and more researchers choose to use the fusion of multiple features to obtain enhanced descriptors and achieve better performance. This algorithm also follows this idea, as shown in Fig. 2. Three features, including HOG, CN, HOI, are respectively used in different parts of the algorithm.

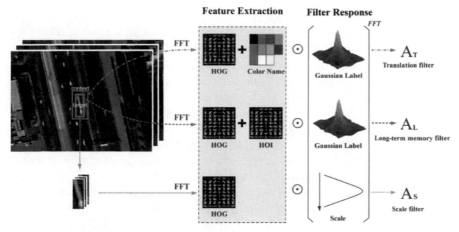

Fig. 2. Detailed construction of the three filters in the algorithm

Histogram of Oriented Gradient (HOG)

HOG is a feature descriptor commonly used in target tracking. The basic idea is that the flexible target is greatly non-rigid, and its color and texture change a lot, so it is difficult to effectively describe the matching of color and texture features and shape template. In contrast, in an image, the directional density distribution of gradients or edges is a good description of the appearance and shape of local objects.

The high-dimensional features are introduced into the framework of correlation filtering. Based on the structure of HOG eigenmatrix, the researchers use sparse representation and dot product processing of eigenmatrix. Because of the sensitivity and insensitivity of the original direction, the pixel level features were mapped, and after normalization, truncation, projection and dot product processing, the 31-dimensional FHOG feature was obtained [8]. In this way, under the condition of fewer model parameters and faster detection speed, the better performance of the algorithm can be guaranteed.

Color Name (CN)

CN is a template class feature made up of 11 colors. The researchers have searched a large number of Google images to obtain a generic color feature matrix, including black, blue, brown, gray, green, orange, pink, purple, red, white and yellow. The feature model is further optimized to make the algorithm highlight the colors with higher frequency and be more representative.

Histogram of Local Intensities (HOI)

The intensity gradients are not robust to the appearance change caused by deformation. Therefore, [9] build local statistical features by referring to the principle of similarity of distribution field scheme, and obtains HOI feature descriptors respectively through the intensity channel and the channel after applying non-parametric local rank transformation to the intensity channel. The robustness is greatly enhanced.

Multi-feature Selection and Fusion

Figure 2 illustrates the feature selection of the three filters in the algorithm. These three filters play different roles in the tracking process, and the learning sample sizes are also different. The translation filter learns information about the target and background. HOG is robust to local deformation and illumination changes, but unstable to image blurring. CN feature compensates the background clutter effectively, and the two are complementary to distinguish positive and negative samples. The scale filter is similar to the long-term memory filter in that it does not extract background information. [10] shows that the addition of HOI features cannot improve the accuracy of scale estimation, so only the addition of HOG and HOI fusion features in the long-term memory filter improves the sensitivity of features to intensity changes.

The kernel is composed of dot product or norm of parameters. Extracting features yields a matrix with C channels connected. Therefore, the fusion feature descriptor is obtained by the single dot product sum calculation of each channel. The correlation filter in the algorithm adopts linear kernels, so the multi-channel version is:

$$\kappa\left(x, x'\right) = x^T x' \tag{1}$$

$$k^{xx'} = F^{-1}\left(\sum_c \hat{x}_c^* \odot \hat{x}_c'\right) \tag{2}$$

Where F^{-1} is the inverse discrete Fourier transform, $\hat{x}_c^*$ represents the complex conjugate form of the sample in the Fourier domain, and $\odot$ represents the multiplication of the elements between matrices.

2.2 Re-detection Mechanism

When the translation filter loses the target in the tracking, the tracker can be reinitialized by the detector. In other words, the detector obtained by off-line training can give the initial position of the target. After tracking failure, the translation filter converts to the online detector (Fig. 3).

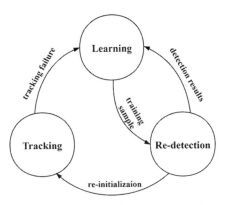

Fig. 3. The process of long-term tracking

The appearance of the target will change over time, and the long-term memory filter needs to be updated.

$$\tilde{x}^t = (1 - \eta)\tilde{x}^{t-1} + \eta x^t \tag{3}$$

$$\tilde{a}^t = (1 - \eta)\tilde{a}^{t-1} + \eta a^t \tag{4}$$

Where $\tilde{x}$ and $\tilde{a}$ are the target size and appearance models in the Fourier domain respectively. t is the current frame and $\eta \in (0, 1)$ is the learning rate.

Meanwhile, as a switch of the re-detection mechanism, the long-term memory filter balances the maximum response value C_{max} with a predetermined threshold T_r.

$$C_{max} = \max(f(z)) \tag{5}$$

Where $f(z)$ is the response map of an image patch z in the Fourier domain.

C_{max} is set to the confidence score. The detector will only be turned on when the re-detection threshold is exceeded. This is to prevent the algorithm from performance sliding window detection on the video every frame, thus reducing the computational cost.

The online SVM classifier is selected as detector in the long-term tracking framework. The training samples of the classifier are determined by the estimated position and scale changes, and the samples are assigned binary tags for training SVM. We refer to the method in [11] and use non-parametric local rank transformation. Therefore, the hyperplane objective function for calculating the SVM detector is:

$$\min_{h} \frac{\lambda}{2} \|h\|^2 + \frac{1}{N} \sum_{i} \rho(h; (v_i, c_i)) \tag{6}$$

$$\rho(h; (v_i, c_i)) = \max\{0, 1 - c\langle h, v\rangle\} \tag{7}$$

Where $\langle h, v\rangle$ represents the inner product of h and v, λ is the regularization coefficient, and (v_i, c_i) is the training set of N samples in a frame.

2.3 PCA Based Dimensionality Reduction

Compared with the current research direction of tracking algorithms, satellite video tracking methods pay more attention to the performance of algorithm speed. In general, the real-time performance of the algorithm should be 20fps/s.

The computational complexity of the tracker is proportional to the feature dimensions. Adaptive dimensionality reduction technology is adopted to reduce the computational cost and increase the frame rate of the algorithm while retaining a large amount of target feature information.

Principal Components analysis (PCA) is the core principle of this dimensionality reduction technique: computing covariance matrix, allowable eigenvector, projection data and obtaining principal component dimension. See [12] for the detailed calculation process.

Considering that the algorithm has three filters and different feature selection, we believe that both HOI and CN features are auxiliary features of HOG, so the focus of dimensionality reduction is not the latter. Therefore, the dimensionality reduction object used in this paper is the feature of the translation filter. HOG feature is reduced from 33 dimensions to 27 dimensions, and CN feature is reduced from 10 dimensions to 2 dimensions. The dimension can be reduced without greatly affecting the accuracy. Similar to the method in [13], a smoothing term is added to ensure the robustness of the projection learning matrix:

$$\varphi_{total}^t = \alpha_t \varphi_{data}^t + \sum_{i=1}^{t-1} \alpha_i \varphi_{smooth}^i \tag{8}$$

Where φ_{data} is the corresponding data term after dimensionality reduction processing of the current frame appearance model, φ_{smooth} is the smoothness term coefficient to ensure the robustness of the projection matrix learning, while both are controlled by the weights α_i.

3   Experimental Results and Analysis

Three groups of video sequences from different locations were selected to perform target tracking experiments to confirm the performance of the proposed algorithm. These meter-resolution satellite videos were captured by the Skybox-1 satellite in Turkey, Las Vegas, and Dubai (Fig. 4). The video parameters are shown in Table 1. The experimental platform is Intel i7-9750H 2.60 GHz CPU and 16 GB RAM, and the development environment is MatlabR2018a.

Satellite video data does not have groundtruth boxes for professional detection, so precision testing cannot be carried out. Therefore, we refer to the tests of most current satellite video tracking studies and use the method of visual comparison of target tracking to analysis the algorithm performance. In order to evaluate the performance of the algorithm more accurately, we selected the relevant sequence of OTB100 [14] set as a supplementary experiments. Because the close-range video dataset has the groundtruth value annotated manually, the precision plot and success plot are used for quantitative analysis.

Precision plot shows the percentage of frames in which the estimated target position is within a certain threshold distance from the groundtruth. And the success plot is defined as:

$$S = \frac{|r_t \cap r_a|}{|r_t \cup r_a|} \tag{9}$$

Where r_t is the tracking bounding box, and r_a is the groundtruth bounding box.

The experiment selects this algorithm (Ours) and its dimensionality reduction version (Ours -PCA) as well as three currently popular methods for testing. Figure 5(a)-Fig. 5(d) respectively show the tracking results of the five algorithms in the Skysat-1 satellite video data. The tracking boxes of algorithm Ours, Ours-PCA, SAMF [15], LCT and CN are

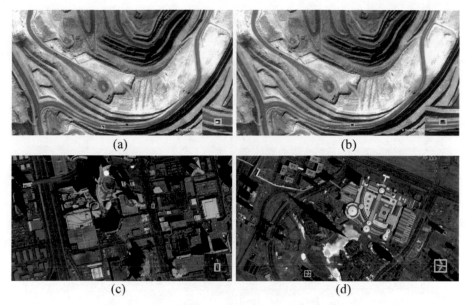

Fig. 4. (a) Turkey1 (b) Turkey2 (c) Las Vegas (d) Dubai

Table 1. Skysat-1 satellite video and tracking target parameters

Video sequence	Frame number	Image size	Target size
Turkey1	1799	1280 × 720	13 × 15
Turkey2	1799	1280 × 720	14 × 11
Las Vegas	1799	1920 × 1080	8 × 20
Dubai	899	1920 × 1080	39 × 41

respectively represented by blue, green, red, yellow and orange. Four frames from each video were extracted for analysis.

As shown in Fig. 5(a) of the tracking of the aircraft in the Dubai video, a target of this size is not difficult for existing advanced trackers. However, when the size of the target is too small and the interference of similar objects is too much, most of the trackers are difficult to ensure the tracking of the target. Both Fig. 5(a2) and Fig. 5(c3) have unrelated targets coming into view. The features of the target itself are untextured, making it difficult to obtain useful information. The comparison algorithms all failed to track.

It can be concluded from Fig. 5 that only the algorithm we proposed can successfully track all the target. The dimensionality reduction version only gradually deviates from the target after frame 1670 in Fig. 5(d4). The frame rate of the algorithm is improved by 37.8%, and real-time tracking above 25fps/s can be guaranteed in all videos (Table 2).

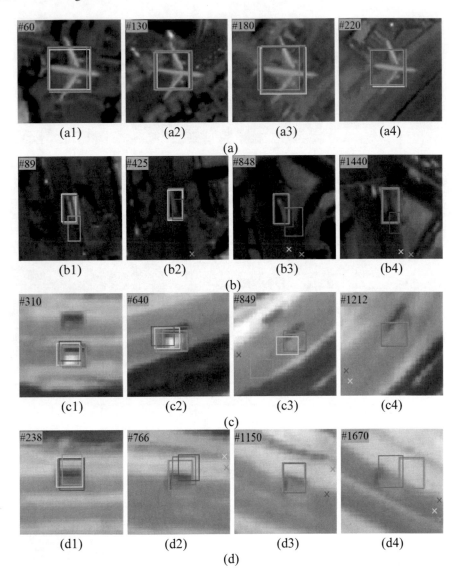

Fig. 5. Satellite video sequences tracking results

Table 2. Dimension difference and frame rate of each tracker

Algorithm	Feature dimension	Re-detection	FPS
Ours	43	Yes	31.58
Ours-PCA	29	Yes	43.53
LCT	47	Yes	17.15
SAMF	43	No	9.49
CN	2	No	230.76

Figure 6 shows the tracking results of five algorithms in the OTB dataset for partial sequences that conform to space-based characteristics. The accuracy of this algorithm is 83.4% and the success rate is 63.2%. The reduced dimension version only lost 0.9% and 0.3% of the two indicators, while the frame rate increased by 15%.

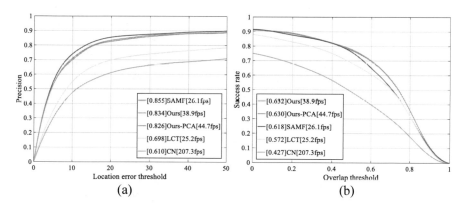

(a)   (b)

Fig. 6. Part of the OTB100 data precision plot and success plot

Overall, the selected features of the proposed algorithm complement each other's shortcomings, and the experimental ressults show that are effective. In addition, the long-term tracking framework solves the loss problem in the tracking process, and improves the robustness by detecting retrieving the target again. More importantly, the introduction of dimensionality reduction technology also doesn't affect the performance loss, allowing the algorithm to maintain high-speed tracking.

4 Conclusion

In this paper, tracking framework and computational complexity are the two directions of algorithm improvement. The gradient histogram, color name and local intensity histogram were selected as features to improve the accuracy of the algorithm. Then, a long-term tracking framework is introduced to provide re-detection for the tracker. Finally, principal component analysis is used to select the most appropriate target and dimension according to the different characteristics of the algorithm. It solves the problems of low precision and poor real-time in space-based remote sensing video of traditional algorithm. The effectiveness of the proposed method is verified by three satellite video experiments. The results show that the comprehensive performance of this method is better than the existing correlation tracking methods.

It is worth noting that the algorithm misses the target many times when the background is similar to the target. It shows that the selection of features needs to be further improved. In addition, the number of filters is proportional to the cost of computation. In the following research, we will make further improvement on whether the scale filter is necessary for the target whose size only changes slightly.

References

1. Lu, H.C., Li, P.X., Wang, D.: Visual object tracking: a survey. Pattern Recogn. Artif. Intell. **31**(1), 61–76 (2018). (in Chinese)
2. Dyer, J.M., McClelland, J.: Paradigm change in earth observation - skybox imaging and SkySat-1. In: Hatton, S. (ed.) Proceedings of the 12th Reinventing Space Conference, pp. 69–89. Springer, Cham (2017). https://doi.org/10.1007/978-3-319-34024-1_5
3. Ahmadi, S.A., Ghorbanian, A., Mohammadzadeh, A.: Moving vehicle detection, tracking and traffic parameter estimation from a satellite video: a perspective on a smarter city. Int. J. Remote Sens. **40**(22), 8379–8394 (2019)
4. Shao, J., Du, B., Wu, C., Zhang, L.: Can we track targets from space? A hybrid kernel correlation filter tracker for satellite video. IEEE Trans. Geosci. Remote Sens. **57**(11), 8719–8731 (2019)
5. Xuan, S., Li, S., Han, M., Wan, X., Xia, G.S.: Object tracking in satellite videos by improved correlation filters with motion estimations. IEEE Trans. Geosci. Remote Sens. **58**(2), 1074–1086 (2019)
6. Dalal, N., Triggs, B.: Histograms of oriented gradients for human detection. In: 2005 IEEE Computer Society Conference on Computer Vision and Pattern Recognition, vol. 1, pp. 886–893 (2005)
7. Van De Weijer, J., Schmid, C., Verbeek, J., Larlus, D.: Learning color names for real-world applications. IEEE Trans. Image Process. **18**(7), 1512–1523 (2009)
8. Felzenszwalb, P.F., Girshick, R.B., McAllester, D., Ramanan, D.: Object detection with discriminatively trained part-based models. IEEE Trans. Pattern Anal. Mach. Intell. **32**(9), 1627–1645 (2010). https://doi.org/10.1109/TPAMI.2009.167
9. Ma, C., Yang, X., Zhang, C., Yang, M.H.: Long-term correlation tracking. In: Proceedings of the IEEE Conference on Computer Vision and Pattern Recognition, pp. 5388–5396 (2015)
10. Ma, C., Huang, J.B., Yang, X., Yang, M.H.: Adaptive correlation filters with long-term and short-term memory for object tracking. Int. J. Comput. Vis. **126**(8), 771–796 (2018)
11. Zhang, J., Ma, S., Sclaroff, S.: MEEM: robust tracking via multiple experts using entropy minimization. In: Fleet, D., Pajdla, T., Schiele, B., Tuytelaars, T. (eds.) ECCV 2014. LNCS, vol. 8694, pp. 188–203. Springer, Cham (2014). https://doi.org/10.1007/978-3-319-10599-4_13
12. Wold, S., Esbensen, K., Geladi, P.: Principal component analysis. Chemom. Intell. Lab. Syst. **2**(1–3), 37–52 (1987)
13. Danelljan, M., Shahbaz Khan, F., Felsberg, M., Van de Weijer, J.: Adaptive color attributes for real-time visual tracking. In: Proceedings of the IEEE Conference on Computer Vision and Pattern Recognition, pp. 1090–1097 (2014)

14. Wu, Y., Lim, J., Yang, M.H.: Object tracking benchmark. IEEE Trans. Pattern Anal. Mach. Intell. **37**(9), 1834–1848 (2015)
15. Li, Y., Zhu, J.: A scale adaptive kernel correlation filter tracker with feature integration. In: Agapito, L., Bronstein, M.M., Rother, C. (eds.) ECCV 2014. LNCS, vol. 8926, pp. 254–265. Springer, Cham (2015). https://doi.org/10.1007/978-3-319-16181-5_18

Fourier Series Fitting of Space Object Orbit Data

Ziwei Zhou[1], Gaojin Wen[1], and Yun Xu[2(✉)]

[1] Beijing Engineering Research Center of Aerial Intelligent Remote Sensing Equipment, Beijing Institute of Space Mechanics and Electricity, Beijing 100192, People's Republic of China
[2] Laboratory of Computational Physics, Institute of Applied Physics and Computational Mathematics, Beijing 100088, People's Republic of China
xu_yun@iapcm.ac.cn

Abstract. Accurate orbit information plays an important role in space and national defense security, such as space object prediction, maneuver detection, collision prevention and so on. Therefore, it is highly necessary to master the characteristics of orbital elements of space objects. In this paper, the Fourier series fitting method is proposed, in which the TLE orbit data is used to analyze the orbit elements of GEO, LEO and HEO. According to the orbit elements of different types of targets, the orbit elements variation rule is approximated by using the fitting method, and the resulting variation function can be used for predictions. The experimental results show that the predictions of this method is promising.

Keyword: Space object · TLE orbit data · Orbital elements · Variation rule

1 Introduction

Since the launch of the first artificial earth satellite Sputnil-1 on October 4, 1957, mankind has embarked on a journey of space exploration. With the increasing number of human space activities, more and more spacecraft are launched into space, creating an increasingly crowded space environment [1]. In the entire space environment, there are a large number of space objects, including a lot of space debris [2]. Orbital research and prediction of space debris plays an important role in current space situation awareness [3].

With the development of space exploration capabilities, the research on space debris has developed from its orbit positioning and orbit prediction to more comprehensive and accurate applications such as collision prevention. For these studies and applications, all rely on the orbit information of space debris. Therefore, more accurately understanding of the orbital elements' variation rule is very helpful for studying the characteristics and variation rule of its orbit. Most of the research on space debris orbits is based on the traditional SGP4/SDP4 algorithm. Through this method, the TLE (Two-Line Elements) track data reported by NORAD (North American Aerospace Defense Command) can be directly converted into position and velocity, so as to further predict the track [4]. However, this result can only reflect the space information of the debris, and cannot

© Springer Nature Singapore Pte Ltd. 2021
Y. Wang and W. Song (Eds.): IGTA 2021, CCIS 1480, pp. 238–251, 2021.
https://doi.org/10.1007/978-981-16-7189-0_19

obtain the results of its orbital elements at various times. However, different orbital types of space debris are affected by different factors, and the variation rule of each track element is also different. Therefore, it is necessary to classify and research the orbital types and summarize them.

Based on the orbital elements of space debris, this paper proposes a fitting method for variation rule of space debris orbital parameters of different orbital types, which can better describe the rule of existing data as a function. When predicting orbital parameters for the next seven days, the accuracy of the prediction results is better using the variation rule function.

The rest is arranged as follows. Related work is discussed in Sect. 2. In Sect. 3, the method of data fitting is introduced. Experiments and analysis are introduced in Sect. 4. Finally, a brief conclusion is drawn in Sect. 5.

2 Related Works

2.1 TLE Orbit Data

TLE provides the average Kepler orbit parameters, which uses a specific method to remove the periodic disturbance term. The research in this paper is based on the TLE published by NORAD [5]. We extract the orbital elements corresponding to each space object from the TLE, and then conduct experiments and analysis. Therefore, a brief introduction to the TLE track report.

TLE is called the two-line elements, which is the orbital data used by the NORAD to determine the position and velocity of space targets. Since two rows of meaningful character strings are used to represent orbital elements, it is called the two-line elements.

Example:

1 25544U 98067A   04236.56031392 .00020137 00000-0 16538-3 0 9993
2 25544   51.6335 344.7760 0007976 126.2523 325.9359 15.70406856328906

The two rows of TLE data represent meanings as shown in Table 1 and 2 below:

Table 1. TLE track describes the format in the first line

Columns	Example	Description
01	1	Line Number
03–07	25544	Satellite Catalog Number
08	U	Elset Classification
10–17	98067A	International Designator

(*continued*)

Table 1. (*continued*)

Columns	Example	Description
19–32	04236.56031392	Element Set Epoch (UTC) *Note: spaces are acceptable in columns 21 & 22
34–43	.00020137	1st Derivative of the Mean Motion with respect to Time
45–52	00000-0	2nd Derivative of the Mean Motion with respect to Time (decimal point assumed)
54–61	16538-3	B* Drag Term
63	0	Element Set Type
65–68	999	Element Number
69	3	Checksum

Table 2. TLE track describes the format in the second line

Columns	Example	Description
01	2	Line Number
03–07	25544	Satellite Catalog Number
09–16	51.6335	Orbit Inclination (degrees)
18–25	344.7760	Right Ascension of Ascending Node (degrees)
27–33	0007976	Eccentricity (decimal point assumed)
35–42	126.2523	Argument of Perigee (degrees)
44–51	325.9359	Mean Anomaly (degrees)
53–63	15.70406856	Mean Motion (revolutions/day)
64–68	32890	Revolution Number at Epoch
69	6	Checksum

2.2 Orbit Elements

Usually the orbital elements refer to the six parameters of Kepler's orbit: orbit semi-major axis a; orbit eccentricity e; mean anomaly angle M; orbit inclination angle i; ascending node right ascension Ω; argument of perigee ω. The first two parameters determine the shape and size of the orbit, the last three parameters determine the position of the orbit surface in space, and the remaining parameter determines the instantaneous position of object on the orbit in space. As shown in the Fig. 1, these six orbital parameters are represented:

The orbital parameters mentioned above have their rate of change in a certain period of time. According to the rate of change, they can be divided into fast variables and slow variables. Fast variables such as the mean anomaly angle M; slow variables include semimajor axis a, eccentricity e, orbital inclination i, ascending node right ascension Ω,

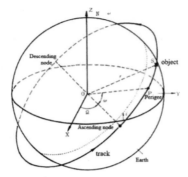

Fig. 1. Schematic diagram of orbital elements

argument of perigee ω, these five orbital elements do not change much in one orbital period, and the change is mainly caused by the perturbation force factor.

This paper mainly studies the characteristics of the variation rule of these five slow variables. Different orbital heights of space debris are subject to different perturbation forces. Therefore, for space debris with different orbital heights, the variation rule of the same orbital parameters are different. In order to more accurately summarize the variation rule of orbital parameters, this paper mainly studies the orbital characteristics of space debris with three types of orbital heights of GEO (Geosynchronous Earth Orbit), LEO (Low Earth Orbit), and HEO (High Earth Orbit), and summarizes the variation rule of orbital elements and applies them to the prediction of their orbital elements.

3   Data Fitting

Curve fitting is a data processing method that uses a continuous curve to approximate or compare the function relationship of discrete points on a plane [6]. It is often used for model prediction, accuracy evaluation and error analysis of discrete data.

3.1   Commonly Used Fitting Algorithms

Commonly used fitting algorithms mainly include the following eight categories:

1) Polynomial fitting, using a polynomial expansion to fit all observation points in a small analysis area containing several analysis grid points to obtain the analysis model of the observation data. The formula can be expressed as:

$$Y(x) = \sum_{i=1}^{N} a_i x^N \quad i = 0, 1, 2, \cdots, N \tag{1}$$

Where a_i is the expansion coefficient; N is the expansion series.

2) Fourier series fitting, mainly use a pair of orthogonal functions to fit the data. Compared with the general method, the expansion parameters have two constant coefficients. The calculation formula can be expressed as:

$$Y(t) = a_0 + \sum_{i=1}^{\infty} (a_i \cos(i\omega t) + b_i \sin(i\omega t)) \quad i = 1, 2, \cdots, N \tag{2}$$

Where a_0 and ω are the constant coefficient; a_i and b_i are coefficient at various levels; N is the expansion series.

3) Exponential fitting, mainly for exponential function curves. For different curve trends, there are two forms of different orders. The formula can be expressed as:

$$Y(x) = ae^{bx}, \ Y(x) = ae^{bx} + ce^{dx} \tag{3}$$

Where a, b, c, d are expansion coefficients.

4) Gaussian fitting, which has a good fitting effect under the curve trend of normal distribution, and its formula can be expressed as:

$$Y(t) = \sum_{i=1}^{N} a_i exp\left(-\left(\frac{x - b_i}{c_i}\right)^2\right) \ i = 1, 2, \cdots, N \tag{4}$$

Where a_i, b_i, c_i are coefficient at various levels; N is the expansion series.

5) Power function fitting, mainly for power function curve, its formula can be expressed as:

$$Y(x) = ax^b + c \tag{5}$$

6) Rational number fitting, mainly based on hyperbolic function model, the formula can be expressed as:

$$Y(x) = \frac{\sum_i P_i x^{N+1-i}}{x^{M+1} + \sum_j Q_j x^{M-j}} \ i = 1, 2, \cdots, N + 1 \ j = 1, 2, \cdots, M \tag{6}$$

Where P_i, Q_j are coefficient at various levels; M and N are the expansion series of the denominator and numerator.

7) Sine fitting, using the sum of multiple-order sine functions for fitting, the formula can be expressed as:

$$Y(x) = \sum_{i=1}^{N} a_i \sin(b_i x + c_i) \ i = 1, 2, \cdots, N \tag{7}$$

Where a_i, b_i, c_i are coefficient at various levels; N is the expansion series.

8) Weibull fitting, which is a special type of S-curve fitting method. The fitting model has a fixed form and a small applicable range. The formula can be expressed as:

$$Y(x) = abx^{b-1} exp\left(-ax^b\right) \tag{8}$$

It can be seen that different fitting methods have different usage ranges due to different curve trends. Aiming at the trend of changes in the orbital elements of space objects or debris, the advantages and disadvantages of the above methods are compared through calculations [7]. The experimental comparison data are given in Sect. 4. Therefore, the Fourier series fitting method is finally chosen to fit the orbital elements, which is also the most consistent with the characteristics of the orbital elements of space debris.

3.2   Fourier Series Fitting

As we all know, the Fourier series is a periodic function, so it can be used to fit any periodic function. There are two ways to express Fourier series. One is the way of trigonometric function. This method is more troublesome for human calculation and analysis, but its mathematical meaning is clear. Computers usually use this when performing Fourier series fitting processing. Fourier series calculation formula is as follows:

$$Y(t) = a_0 + \sum_{i=1}^{\infty} (a_i \cos(i\omega t) + b_i \sin(i\omega t)) \; i = 1, 2, \cdots, N \tag{9}$$

Fourier series is a superposition sequence, in a time interval, the frequency of sine and cosine is an integer multiple of the time interval. The constant is called the average value of the series, and the pairing of sine and cosine at a specific frequency is called the harmonic of the series. The Fourier series can be tailored to any period length. As the number of harmonics increases, the Fourier series converges to any smooth periodic function [8].

4   Experiment and Analysis

4.1   Classification Selection of Experimental Samples

This paper mainly studies the orbital elements of three types of space objects or debris: GEO, LEO, and HEO. Select specific objects from these three types for statistics and research of orbital elements. The following is the basis for the classification of different types of tracks:

$$GEO : 0.99 \leq Mean\,Motion \leq 1.01 \,\&\&\, Eccentricity < 0.01$$

$$LEO : Mean\,Motion > 11.25 \,\&\&\, Eccentricity < 0.25$$

$$HEO : Eccentricity > 0.25$$

Where *Mean Motion* epresents the average number of motion circles of the orbit of the space object or debris; *Eccentricity* represents the orbit eccentricity of the space object or debris.

This paper selects three types of orbit objects: GEO, LEO, and HEO, and selects data from January 1, 2018 to December 31, 2020, for a total of three years for processing and analysis.

4.2   Data Pre-processing

After selecting the number of a space object or debris, we collect statistics on the TLE corresponding to the ID number in three years. It should be noted that these space objects or debris will not be observed every day for three years. There will be days without observation data. At this time, we set the orbital element corresponding to that day to 0. Similarly, there may be cases where there are multiple sets of TLEs for the observed target in a day. In this case, we usually only select the last set of TLEs as the day's orbital data.

Take the GEO object as an example, select GEO ID = 31577, as shown in Fig. 2(a)–(e) below, which shows the target's statistical results of orbital semi-major axis, orbital eccentricity, orbital inclination, ascending node right ascension and argument of perigee for the three years (2018–2020).

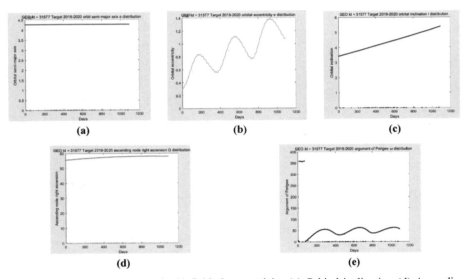

Fig. 2. (a) Orbit semi-major axis (b) Orbital eccentricity (c) Orbital inclination (d) Ascending node right ascension (e) Argument of Perigee

Then, in order to explore the trend of variation in the three-year period, these zero points, where there is no observation data, are interpolated to supplement the complete three-year data to ensure the completeness of the three-year data. After statistics and observations of existing data, we found that except for some special locations, these orbital elements generally show a linear trend (linear increase or linear decrease), so linear interpolation can be used. As shown in Fig. 3(a)–(e) below, it is the result of the interpolation of each orbital element:

For the argument of argument of perigee, it is a change modulo 360°, so it is spliced into continuous changes and then interpolated:

In this link, it should be noted that the appearance of 0 usually has two forms: one is the appearance of a single zero, and the other is the appearance of multiple consecutive zeros. For these two cases, the first case can directly use the average value of the day before and after zero to represent the supplementary data of that day. If there are multiple consecutive zeros, we need to find the first non-zero and last non-zero position where consecutive zeros appear and corresponding data. Perform linear interpolation based on these two known data to supplement the data corresponding to the positions of these consecutive zeros. Supplementing all the data for three years will lay a foundation for

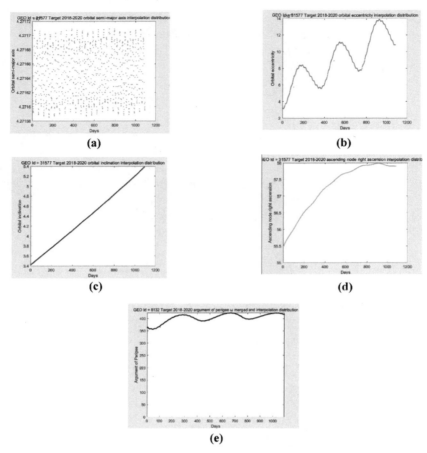

Fig. 3. (a) Orbit semi-major axis interpolation (b) Orbital eccentricity interpolation (c) Orbital inclination interpolation (d) Ascending node right ascension interpolation (e) Argument of Perigee interpolation

the subsequent use of Fourier series to fit its variation rule, so that the fitting result is more in line with the variation rule of the orbital elements of space objects or debris.

4.3 Evaluation Criteria of Fitting Results

The following four evaluation indicators are used to measure the quality of the fitting results [9]:

a) SSE (sum of squares due to error): The sum of squares due to error. The closer the value is to 0, the better the effect of model selection and fitting. This parameter counts the sum of the squares of the errors between the fitted data or predicted data and the corresponding points of the original data. The calculation formula is as follows:

$$SSE = \sum \left(Y_{actual} - Y_{predict} \right)^2$$

b) RMSE (root mean square error): Root mean square error. The closer the value is to 0, the better the curve fitting effect. This statistical parameter, also called the standard deviation of the regression coefficient, is the square root of the MSE, and the calculation formula is as follows:

$$RMSE = \sqrt{MSE} = \sqrt{\frac{1}{n}\sum\left(Y_{actual} - Y_{predict}\right)^2}$$

Among them, MSE is the mean value of the corresponding error sum of squares, $MSE = SSE/n$.

The above two evaluation indicators are based on the error between the original data and the fitted data, that is, the point-to-point error. The following two evaluation indicators are developed relative to the average value of the original data.

c) R-square (coefficient of determination): coefficient of determination, the calculation formula is as follows:

$$R^2 = 1 - \frac{\sum\left(Y_{actual} - Y_{predict}\right)^2}{\sum\left(Y_{actual} - Y_{mean}\right)^2}$$

The denominator can be understood as the degree of dispersion of the original data, and the numerator is the error between the fitted data or the predicted data and the original data. Dividing the two can eliminate the influence of the degree of dispersion of the original data. The closer the coefficient is to 1, the stronger the ability of the equation's variables to solve y and the better the model's fit to the data.

d) Adjusted R-square (degree-of-freedom adjusted coefficient of determination): The adjusted coefficient of determination. The closer the value is to 1, the better the curve fitting effect. Calculated as follows:

$$R^2_{adjusted} = 1 - \frac{\left(1 - R^2\right)(n - 1)}{n - p - 1}$$

Among them, n is the number of samples and p is the number of features. Compared with the coefficient of determination, the correction coefficient of determination eliminates the influence of the number of samples and the number of features.

4.4 Data Fitting

The fitting of each orbit element of different orbit types needs to determine two points. One is the selection of fitting models for each orbital element of different orbit types. The second is that after determining the selected fitting model, it is necessary to determine the fitting model series suitable for the orbital element.

For the space debris of GEO orbit type, select the space object with ID = 31577 for simulation selection and confirmation. The five orbital elements of the target are processed with different fitting models, and the fitting results are compared, so as to select a fitting model that can more accurately describe the variation rule of orbital elements in GEO type. Taking the fitting of the semi-major axis of its orbit as an example, the comparison results are shown in the following table:

Table 3. GEO ID = 31577 Comparison of the results of different models fitting the semi-major axis of the orbit

Fitting method	SSE	RMSE	R-square	Adjusted R-square	Series
Polynomial	150.2514	0.3723	0.0015	−0.0049	7
Fourier	6.6144	0.0784	0.9560	0.9954	3
Exponential	150.2912	0.3717	0.0013	−0.0015	2
Gauss	137.5248	0.3584	0.0860	0.0689	7
Power	150.2904	0.3715	0.0013	−5.7011e-04	2
Rational	–	–	–	–	–
Sine	1.3633e+08	354.7926	−9.0593e+05	−9.1263e+05	3
Weibull	1.9926e+12	4.2756+04	−1.3241e+10	−1.3254e+10	–

It can be seen from Table 3 above that according to the evaluation criteria of the fitting results, the closer the R-square and Adjusted R-square results are to 1, the better the result; and the closer the RMSE is to 0, the better the result. Therefore, it can be seen that in all fitting models among them, the Fourier series fitting method has the best results, so this model is selected.

In the same way, the selection of fitting models for orbital eccentricity, orbital inclination, right ascension of ascending node, and argument of perigee adopts the same method for the semi-major axis of the orbit. For the space objects of GEO orbits, it is more appropriate to use the Fourier series fitting method for the variation rule of these five orbital elements.

The LEO and HEO orbital space debris are processed by the same method of analyzing GEO types. The comparison experiment results show that the Fourier series fitting model is better for the orbital elements of LEO and HEO type space objects.

In the use of Fourier series to fit data, it is necessary to consider the series of the model according to actual needs to avoid problems such as under-fitting and over-fitting in the calculation time process. Therefore, in the selection of the fitting series, the fitting results need to be considered, and the actual significance of the fitting results must also be considered. Perform 2–9-order Fourier fitting calculations on each orbit parameter of the three orbit types of targets, and compare the fitting results, as shown in the following table:

It can be seen from Table 4 that for the orbit semi-major axis of the GEO target, considering the calculation complexity, fitting effect and practical significance [10], it is finally determined that the third-order Fourier fitting is better. In the same way, we determined the Fourier fitting series of the remaining four orbital elements, and fitted the four-order Fourier for the orbital eccentricity, orbital inclination, right ascension of the ascending node, and argument of perigee.

Using the same processing and comparison method, the Fourier series fitting is performed on LEO and HEO targets, and the fitting series of the five orbital elements of these two types of targets are determined respectively. Select LEO ID = 40960, HEO ID = 15680, and the final fitting series are shown in the following Table 5:

Table 4. Fitting results of orbit semi-major axis with GEO ID = 31577 under different orders

Fitting evaluation index	Series							
	1	2	3	4	5	6	7	8
SSE	8.3377	6.7191	5.9738	5.9717	5.9696	5.9502	6.6144	6.6082
RMSE	0.0875	0.0787	0.0742	0.0743	0.0743	0.0743	0.0784	0.0784
R-s	0.9446	0.9553	0.9603	0.9603	0.9603	0.9605	0.9560	0.9561
A R-s	0.9444	0.9551	0.9600	0.9600	0.9599	0.9600	0.9554	0.9554

Table 5. Determination of fourier fitting series for each track element of different types of targets

	GEO	LEO	HEO
Orbital semi-major axis	3	7	5
Orbital eccentricity	4	6	6
Orbital inclination	4	7	7
Ascending node right ascension	4	–	–
Argument of Perigee	4	–	–

The ascending node right ascension and argument of perigee of the LEO and HEO orbits are relatively special due to their special variation. Here is a statistical diagram of the two orbital elements of the LEO target, as shown in Fig. 4(a)–(b) below.

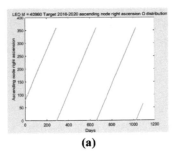

(a)

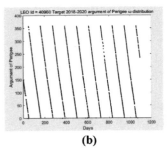

(b)

Fig. 4. (a) Ascending node right ascension interpolation (b) Argument of Perigee interpolation

When fitting these two orbital elements, the Fourier series fitting method is not used, and the linear representation can be used with a period of 360°. Whenever it increases from 0° to 360°, it restarts from 0° in the same increment. Or the decreasing trend linearly increases or decreases.

4.5   Predicting Results

This paper predicts the five orbital elements of the three types of targets, GEO, LEO, and HEO, and predicts the data for the next seven days by fitting seven days of known data. Taking GEO ID = 31577 as an example, this paper selects the data of the 900th to 906th days in three years as the original data of the fitting, and predicts the data of 906th to 912th days. As shown in the figure below, the triangle corresponds to these seven days. The actual value obtained from the TLE data, and the diamond shape is the predicted result obtained through prediction.

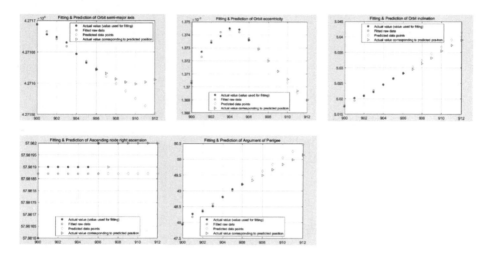

Taking HEO ID = 15680 as an example, the prediction results of its various orbital elements:

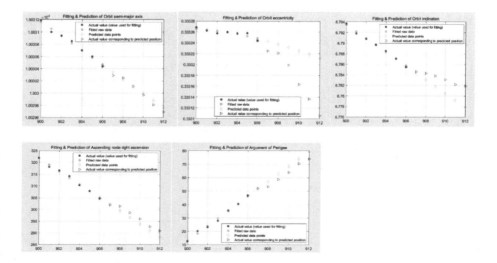

Taking LEO ID = 40960 as an example, the prediction results of its various orbital elements:

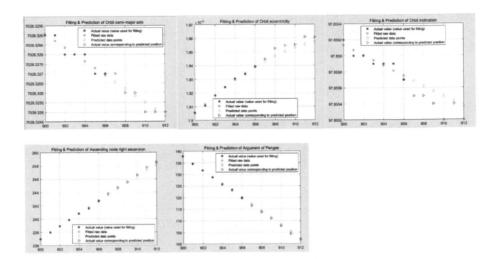

According to the prediction results of GEO, LEO, and HEO, it can be seen that the accuracy of predicting orbit elements is better. Therefore, the fitted variation rule function can be used to predict orbit elements within seven days.

5   Conclusion

This paper mainly analyzes the orbital elements of the three orbital types of space debris, and proposes a fitting method for different orbital parameter variation rule of the three types of orbital space debris: GEO, LEO and HEO. The main fitting method uses Fourier series fitting, which determines the order of Fourier series fitting for different targets. For the prediction of space debris orbital requirements, the variation rule function obtained by fitting the known seven-day orbit elements is used to predict the orbit elements of the next seven days. The experimental results prove that the variation rule function obtained by the fitting can be more accurate. Finally the orbital elements are predicted.

Acknowledgment. The authors gratefully acknowledge the support to this work from all our colleagues in Beijing Engineering Research Center of Aerial Intelligent Remote Sensing Equipment. This work was supported by National Natural Science Foundation of China No. 11772067.

References

1. Li, B.: Research on several key issues of rapid and precise orbit determination and prediction of space debris (2017)
2. Zhang, P.: A method for determining the initial orbit of a space target using only angular observations (2017)

3. Song, B.: Development of U.S. space-based situational awareness system. Int. Space (2015)
4. Liu, W., Miao, Y.: Forecast reliability analysis of SGP4/SDP4 model. Astron. Res. Technol. **2**, 128–131 (2011)
5. Diao, N., Liu, J., Sun, C., et al.: Satellite orbit calculation based on SGP4 model. In: The Fifth Young Marine Scientists Forum of the Chinese Ocean Society and the First Symposium of the Young Marine Science Foundation of the State Oceanic Administration. China Oceanographic Society; First National Oceanic Administration (2011)
6. Shi, L., Nie, X., Ji, M., et al.: List curve fitting based on Matlab curve fitting toolbox. New Technol. New Process (7), 39–41 (2007)
7. Qu, W., Liu, H., Qin, C., Liu, J.: Antenna tracking accuracy evaluation method based on MATLAB Fourier curve fitting. Electron. Meas. Technol. **43**(12), 91–95 (2020)
8. Brooks, E.B., Thomas, V.A., Wynne, R.H., et al.: Fitting the multitemporal curve: a fourier series approach to the missing data problem in remote sensing analysis. IEEE Trans. Geosci. Remote Sens. **50**(9), 3340–3353 (2012)
9. Wei, Y., Zhao, F.: Evaluation criteria for curve best fit. Surv. Mapp. Sci. (1), 195–196 (2010)
10. Qu, W., Li, Y., Li, Y., et al.: Two-antenna signal endpoint identification method using Fourier fitting. J. Ordnance Equip. Eng. **241**(8), 98–102 (2018)

Mapping Methods in Teleoperation of the Mars Rover

Jia Wang[1], Tianyi Yu[1], Junjie Yuan[2], Lichun Li[1], Man Peng[3], Fan Wu[1], Shiying Liu[1], Wenhui Wan[3(✉)], and Ximing He[1]

[1] Beijing Areospace Control Center, Beijing 100094, China
heximing15@nudt.edu.cn
[2] Key Laboratory of Spacecraft In-Orbit Fault Diagnosis and Maintenance, Xi'an 710043, China
[3] State Key Laboratory of Remote Sensing Science, Aerospace Information Research Institute, Chinese Academy of Sciences, Beijing 100101, China
{pengman,wanwh}@radi.ac.cn

Abstract. Mapping methods based on multi-source images are the core technology in the teleoperation of Mars rovers. This study discusses the related unique control characteristics and introduces a teleoperation control mode of the Mars rover based on the "perception, detection, movement" patrol cycle. The multi-scale landing site mapping method based on multi-source data (orbit/descent/ground images) supporting the teleoperation control is described in detail, and its applications in lunar exploration missions are demonstrated. The wide baseline mapping method aimed at mapping large (e.g., mountain peaks) and long-distance targets after landing is proposed, with relevant experiments conducted by the Yutu-2 rover. The ranging error of the panoramic camera in 560 m range is about 4.1 m, and the accuracy is about 0.73%. The wide baseline model was experimentally confirmed to effectively guide task implementation with a high-precision acquisition of the long baseline stereo, laying the foundation for high-precision terrain applications.

Keywords: Mars rover · Teleoperation · Mapping · Multi-source images · Wide baseline

1 Introduction

Mars is not only located near the Earth but also the most Earth-like planetary neighbour [1, 2]. Therefore, the exploration and study of Mars will help to further understand the formation and evolution of Earth and the solar system and predict the future trend of the Earth [3].

The exploration of Mars is significantly more difficult compared with lunar exploration because the distance from Earth to Mars is much larger than the distance from Earth to the Moon. Therefore, up to now, human exploration of Mars has only involved three kinds of methods: overflight, orbitally-based-remote sensing, and descend-land and patrol exploration [4]. The implementation of these Mars missions, where some of the probes are still in orbit, has obtained a large number of images of the surface of Mars. These images have been widely used in Mars scientific research during and

© Springer Nature Singapore Pte Ltd. 2021
Y. Wang and W. Song (Eds.): IGTA 2021, CCIS 1480, pp. 252–264, 2021.
https://doi.org/10.1007/978-981-16-7189-0_20

after the missions; they have enabled important engineering progress and a large number of scientific research achievements. In 2015, the National Aeronautics and Space Administration (NASA) held a press conference to announce the indirect evidence of the existence of liquid water on Mars [5], which once again stimulated expectations and the enthusiasm for Mars exploration.

Compared with overflight and orbitally-based-remote sensing, descend-land and patrol exploration can explore the surface of Mars in situ at a higher resolution, which is an important means for surface exploration and scientific research of Earth-like planets [6]. High-precision mapping based on multi-source images is crucial for the successful completion of a variety of deep space mission operations and scientific investigations [6]. By combining orbital/descent images and ground data, different scales of landing site mapping and topographic analysis can be realized. Herein, we present some of the results of these analyses collected by the teleoperation team of the Beijing Area Control Center, in collaboration with the Planetary Remote Sensing team of the State Key Laboratory of Remote Sensing Science, Institute of Remote Sensing and Digital Earth, Chinese Academy of Sciences. The multi-scale mapping mode discussed here effectively supports the teleoperation control of the rover during each planning stage.

The remainder of this study is organized as follows: Sect. 2 introduces the teleoperation mode of the Mars rover based on the predesigned "perception, detection and movement" cycle. In Sect. 3, we elaborate on the mapping method based on the collection of multi-scale multi-source data. Furthermore, we report on the mapping results based on the wide baseline of the Yutu-2 rover using panoramic cameras. Conclusions and future developments are finally discussed in Scct. 4.

2 Mars Rover Teleoperation Control Mode

After landing on the Mars surface, the rover is required to adapt to the Martian surface environment by performing patrol activities in the landing area and sending the detection data back to Earth [7]. The specific functions include [8]: cooperating with the landing platform to complete the separation process and safely reach the Mars surface; navigate on the landing area using the mobile abilities such as moving forward and backward, steering, climbing, obstacle surmounting; environmental perception, attitude determination, relative positioning, path planning, motion control, etc.; carrying and placing payload to specifically detected targets; establishing measurement, control, and communication links with data management and transmission capabilities.

As opposed to the lunar high-vacuum and rigid environment, the Mars surface exhibits time-varying characteristics of the atmosphere and wind fields, while possessing a higher surface rock coverage [9]. Due to long-term weathering, the outer surface is soft, and the terrain can be deceptive. Moreover, the solar radiation intensity on the Mars surface is approximately 20% of that on the lunar surface. If there is dust, solar radiation will be further mitigated, and thus, the solar cell power generation of the rover will be severely hindered [10]. These factors have brought considerable challenges to the energy generation functionality of the Mars rover. Considering the limited tracking, telemetry, and command (TT&C) resources of the mission, along with the complex and changeable Mars surface environment and large communication delay, the Mars rover should improve its autonomous performance in energy management, mission planning, thermal control management, and communication management [11].

In the Mars exploration missions that have been carried out as of now, the rover driving operations were not fully automatically; hence, it had to be operated remotely from the ground control center. In the ground control center, the operators integrated the surface environment of Mars, autonomous ability of the rover, the demand for scientific exploration, measurement and control conditions and other factors, to reasonably plan the routes, scientific and engineering scopes of the rover, and finally converted them into control instructions [12].

The most important basic data of the Mars rover teleoperation planning is information regarding the Mars surface terrain [13]. The unique planning characteristics of each stage require specific terrain data, mainly in terms of coverage area size and resolution. Specifically, a primary requirement for the strategic layout of the overall planning is a sufficiently large coverage of terrain features, i.e., in the order of 100 km. The cycle planning is usually formulated after landing, and its scope is required to accurately cover the landing site (generally in the order of kilometers or hundreds of meters), while considering the trajectory factor of the descent process. Unit planning is mainly employed for a specific rover movement; hence, the first requirement is a sufficiently high resolution of the terrain features (generally in the centimeter scale).

3   Multi-scale Mapping Mode Based on Multi-source Data

This section discusses the multi-scale mapping method with multi-source data, namely orbit/descent/ground images taken from the rover (Fig. 1). The process is divided into four stages: (1) Before launch, where prior data of Mars orbiters are used to analyze the geological background and identify a large range of landing area terrain. (2) In the parking orbit, the high-resolution camera captures multi-orbit images from the landing area, which are used to determine the surface terrain of the landing area accurately. These data are primarily used for overall planning. (3) During the entry/descent/landing (EDL) process, specific imaging equipment on the lander/rover probe will be turned on; these equipment will send a sequence of images of the Mars surface. By integrating the relevant data, a more accurate terrain imaging of the landing site can be generated for future cycle planning. (4) During the surface operations, the navigation terrain and obstacle avoidance cameras are used periodically to image the forward direction at each navigation point. Local digital elevation models (DEMs) will be routinely produced, which combined with ground-level topographic analysis, will support the waypoint-to-waypoint path planning. The mapping methods of (1) and (2) are similar. The data source is the orbiter images, and the imaging equipment is the linear array camera. The stereo coverage is realized by the side swing of the probe.

3.1   Large Area Mapping Using Orbit Images Based on the Two-Stage Method

Large area image mapping involves an enormous number of images, which combined with the refinement of each image file, will lead to low efficiency and accuracy. In addition, in the process of image mapping, a DEM is needed to help eliminate the projection

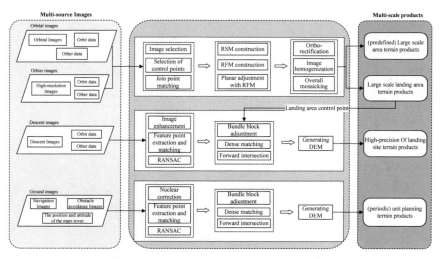

Fig. 1. Multi-scale mapping method based on multi-source data.

difference caused by the terrain, and therefore, alleviating the geometric inconsistencies between the images and the control data is also required. To guarantee both processing efficiency and mapping precision, a "two-stage" high-precision geometric processing scheme for large area imaging is adopted, in which the large areas are partitioned into several subareas processed in parallel [14].

Initially, the images in the entire predefined landing area are partitioned based on the established geometric image model, and planar block adjustments are applied in each subarea, so as to alleviate the geometric deviations between the images in the partition as well as the geometric inconsistencies between the images and the control data. On this basis, orthophoto correction and image mosaic functions are implemented. Owing to the resolution limitation of the reference source, some positional inconsistencies of the digital orthophoto map (DOM) mosaics between neighboring subareas persisted. Therefore, a thin plate function (TPS) model-based image registration is applied to the generated DOM mosaics of each subarea. To maintain grayscale and contrast homogeneity, a histogram matching-based grayscale balancing method is applied to every DOM. Finally, a seamless DOM product of the entire predefined landing area is generated via mosaicking. The technical process steps are illustrated in Fig. 2 [14, 15].

During the preparations of the Chang'e-5 (CE-5), we utilized this method to generate a seamless DOM mosaic of the entire predefined landing area. The generated radiometrically homogeneous and geometrically seamless DOM mosaic is illustrated in Fig. 3 (zoomed-out view). The control point root mean square (RMS) errors were approximately one SLDEM2015 grid cell in size, signifying that the produced DOM had been registered to SLDEM2015 with high precision. The final DOM mosaic image contained 224721 columns and 44945 rows with a ground resolution of 1.5 m, covering a surface of 413.8 km × 121.4 km [14]. This DOM played a vastly important role in the specific selection of the landing site, along with the evaluation of the geological environment and the analysis of the terrain/geomorphic characteristics specific to the CE-5 mission.

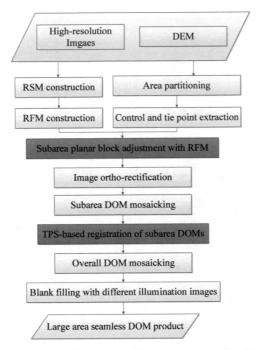

Fig. 2. Process flowchart of large-area high-resolution seamless DOM generation.

It was also used as one of the base maps for visual localization of the CE-5 lander using descent images [16].

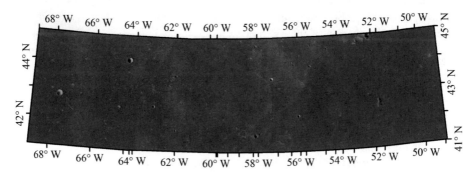

Fig. 3. High-resolution LROC NAC DOM mosaic of the CE-5 landing site region in Lambert conformal conic projection.

The design index of the high-resolution camera of Tianwen-1 probe can achieve an image resolution that exceeds 0.6 m at an orbit height of 300 km near the Mars point of landing [17]. By employing the image data from a high-resolution camera, the landing area can be more accurately mapped; the resolution is expected to surpass the meter-level scale.

3.2 Landing Site Mapping Using Descent Sequence Images

Compared with the orbit images, the landing area terrain product generated by the descent sequence images has a higher resolution and more accurate coverage. During the descent process, the landing area was continually imaged with the descent camera, which obtained sequence images with a higher resolution than that of the orbiter images. Such imaging can generate more precise terrain features for detection cycle planning [18]. Owing to the presence of the atmosphere and wind during landing, image enhancement based on haze removal is carried out before image matching; the main technical scheme of this process is shown in Fig. 4:

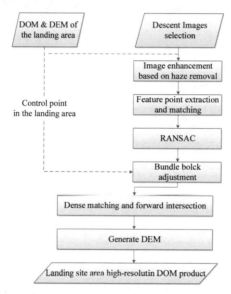

Fig. 4. Mapping process of descent sequence images.

In CE-5, 87 descent images capturing 2.8 km of the lunar surface were selected to map the landing area (Fig. 5). The corresponding DOM resolution was 0.5 m, and the coverage area was 2.5 km × 1.8 km. Because the dynamic descent trajectory of the detector was inclined, there was an obvious intersection angle between the sequence images; therefore, we generated the high-resolution DEM in a synchronous manner. Using the descent sequence images and the predefined DOM produced from the LROC NAC images as the base map, we achieved a precise and prompt landing localization [16]. The high-precision mapping products of the landing area provided the key basic information for the rapid analysis of the sampling area, as well as the rapid positioning of the sampling, lifting, and canning, and the subsequent ascension sequence.

During the EDL process, the descent camera will send a sequence of images of the Mars surface. Using similar technology, high-resolution and accurate range mapping of the landing site can be obtained, with a resolution quality exceeding the meter scale. This data can also be used for post-landing cycle planning. In addition, feature matching and positioning based on descent images and high-resolution images of the orbiter are

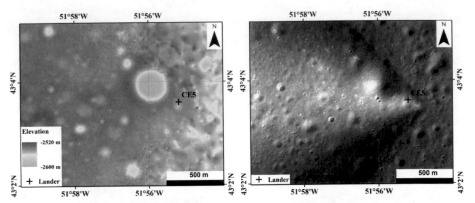

Fig. 5. CE-5 landing site DEM and DOM (0.05-m resolution) generated from descent images. DEM on the left, and DOM on the right.

important for the Mars rover to carry out follow-up scientific exploration activities. Antenna pointing control, attitude determination and other behaviors require accurate landing site longitude, latitude, and elevation information [7].

3.3 3D Terrain Reconstruction Based on Image Fusion

Terrain reconstruction is not only an important tool to understand the Mars environment, but also an essential component in the autonomous navigation of the rover. Additionally, realistic terrain reconstruction helps enhance the immersion for the teleoperators, effectively improving their efficiency.

Before the rover moves, the teleoperation center adjusts the pitch and yaw drives of the mast to enable the navigation camera (Navcam) to capture stereo images at different angles. The teleoperators then use dense matching technology to process the downlinked images. In view of the complex terrain environment and the rugged terrain of the landing area, a pair of obstacle avoidance cameras is placed at the front and rear of the rover, which can realize obstacle detection in any direction [19]. Therefore, the image fusion processing method can be adopted, and the terrain data from the hazard avoidance camera (Hazcam) are used to fill in the blind spot in the terrain data from the Navcam. The specific implementation of this process is shown in Fig. 6.

Figure 7 shows the path planning of Yutu-2 rover to select the path to the "dormant" point. Figure 7(a) shows the DOM at this waypoint, which was automatically generated from 16 pairs of Navcam images captured at a fixed pitch angle, and Fig. 7(b) is the left Hazcam image that shows a large pit in the area around the front left wheel of the rover. Figure 7(c) is the DOM generated from Hazcam images. Considering the low resolution of Hazcam image in the far range, we only cut and used the DOM within the range of 3.5 m. Figure 7(d) shows the merged Navcam and Hazcam DOM, within which the Hazcam DOM is indicated by the red oval.

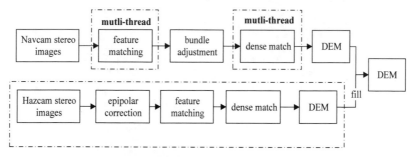

Fig. 6. Flowchart of terrain reconstruction based on image fusion

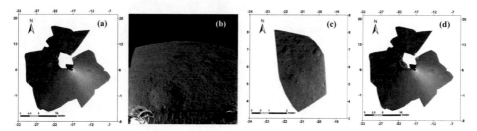

Fig. 7. The DOMs and Hazcam image: (a) Automatically generated DOM from Navcam stereo images; (b) Left Hazcam image; (c) Automatically generated DOM from Hazcam stereo images; (d) DOM after merging.

3.4 Wide Baseline Mapping Using Multi-site Images

Among the numerous scientific equipment carried on the Mars rover, the Navcams and Hazcams are the most important instruments for mapping. However, due to the short baseline, their terrain products are incapable of supporting medium - and long-term path planning. By establishing a wide baseline using multi-sites, the effective mapping range of the rover can be extended from tens to hundreds of meters [20].

According to the principle of photogrammetry and error-propagation derivation, the range error can be calculated as [21, 22]:

$$\sigma_Y^2 = \frac{Y^2 B^2}{8 b_{navcam}^2 f_{navcam}^2} \sigma_p^2 + \left(\frac{Y^2}{B f_{pancam}}\right)^2 \sigma_p^2 \tag{1}$$

where σ_Y is the standard error of range Y, σ_p is the parallax measurement error, B is the wide baseline, and b_{navcam} is the baseline of the navigation terrain cameras. Note that the first term is the baseline error, and the second term represents the parallax measurement error.

By substituting the camera parameters [23, 24] of Yutu-2 rover into Eq. (1), we can generate the range measurement errors of the targets at different distances from the camera (100 to 700 m) under different wide baselines (Fig. 8). From the aforementioned discussion, it can be inferred that the mapping errors do not change monotonically with respect to the baselines, and that an optimal baseline exists for a given distance to the target. Table 1 presents the optimal baselines for targets at different distances.

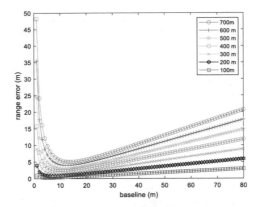

Fig. 8. Range measurement error of different Pancam wide baselines for the Yutu-2 rover.

Table 1. Optimal baseline for targets at different distances based on theoretical analysis. (unit: m).

Target distance	100	200	300	400	500	600	700
Optimal baseline	5.1	7.2	9.0	10.2	1.4	12.6	13.5
Range error	0.270	0.763	1.402	2.159	3.018	3.966	4.999

To verify the validity of the wide baseline mapping method, we designed the corresponding data acquisition strategy and implemented it on the 4th and 5th lunar day. The captured stereo images and their detailed views are shown in Figs. 9 and 10 and Table 2.

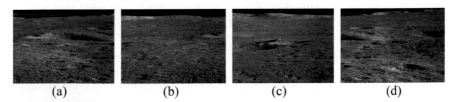

(a)   (b)   (c)   (d)

Fig. 9. Four left Pancam images captured at site 1 (on the 4th lunar day).

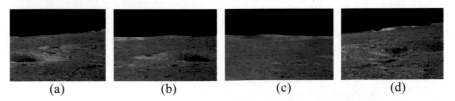

(a)   (b)   (c)   (d)

Fig. 10. Four left Pancam images captured at site 2 (on the 5th lunar day).

Table 2. Imaging information of the Pancam images.

Stereo image No.	Position of Yutu-2 (m)			Attitude of the Yutu-2 (°)			Attitude of the mast (°)		
	x	y	z	Roll	Yaw	Pitch	Roll	Yaw	Pitch
9-a	33.52	−112.1	0.82	1.11	166.02	2.52	−0.066	−43.11	−6.86
9-b								−30.15	
9-c								−17.12	
9-d								−56.12	
10-a	36.71	−119.2	1.09	0.21	164.56	6.86	−0.088	−43.11	−6.92
10-b								−30.15	
10-c								−17.12	
10-d								−56.12	

We used the SURF-64 feature descriptor process and applied a distance-ratio measure (the ratio of Euclidean distance of the closest neighbor to that of the second-closest neighbor) for matching [21]. As a result, SURF provided a sufficient number of matched points for close-and-medium-range terrain, but only a few for far-range terrain. By combining the matched points from both the Förstner and the SURF matching processes, we were able to obtain a sufficient number of matched points for close-medium- and far-range terrain scenarios.

The left side of Fig. 11 illustrates the partial feature-point matching of a wide baseline stereo pair taken by the Yutu-2 rover. If a line shows significant differences in orientation and length compared to the neighboring lines, it is a mismatch. To eliminate these mismatches, we computed the fundamental matrix between the wide baseline image pair using the RANSAC [25] procedure. Under this iterative procedure, mismatches were automatically eliminated as outliers. The right side of Fig. 11 illustrates the results of the feature-point matching process after outlier elimination.

Therefore, only the wide baseline tie points were selected from the corrected matched points. By drawing a grid (e.g., 3 × 3) on the image and selecting the one-tie point having the highest correlation coefficient (or distance ratio) in each patch, we can obtain evenly distributed wide baseline tie points that are suited for the subsequent bundle adjustment. Finally, the three-dimensional coordinates of the matching points were calculated using the corrected exterior orientation parameters (EOPs).

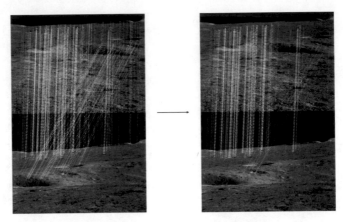

Fig. 11. Matched feature points. Left – before, and right – after outlier elimination.

By selecting some obvious large impact craters from the image, we calculated the distance between the crater and the rover based on the feature points of the crater. According to the calculations, the distance between the crater and the rover was approximately 560.1 m. Simultaneously, we measured the distance between the two sites and the impact crater in the LRO NAC images, which were 566.5 m and 558.1 m, respectively (Fig. 12). The results indicated that the ranging error and accuracy were approximately 4.1 m and 0.73%, respectively, which were consistent with the derived ranging error.

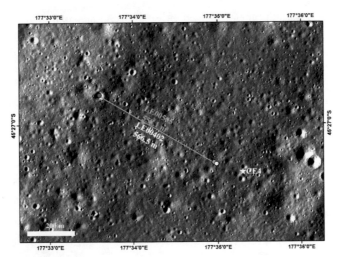

Fig. 12. Distance between the impact crater and the Yutu-2 rover measured on the LRO NAC base map, which is a high resolution (0.9 m/pixel) DOM [26] generated from Lunar Reconnaissance Orbiter Camera Narrow Angle Camera imagery [27].

4   Conclusions

In this study, the remote operation control mode of Mars rover based on three-layer planning was introduced. We then discussed the method of multi-scale and high precision 2D and 3D mapping of the Mars surface using multi-source data in each planning stage and verified the method with Chang'E data. At present, the Mars exploration mission of Tianwen-1 is planned to realize orbiting, landing and surface inspection in a signal mission. This demands superior remote operation and environment perception performance. The method proposed here is expected to provide technical reference for remote operation detection of Mars rovers.

Future work that might contribute to the field of Mars remote sensing mapping includes: global surface mapping of Mars based on multi-source, multi-coverage remote sensing data; improving the Mars global control network; mass and automatic remote sensing data processing and information mining; real-time long-distance navigation and mapping of the routes of the Mars rover.

Acknowledgments. This study was supported by the National Natural Science Foundation of China (grant Nos. 41771488, 11773004).

References

1. Wan, W.X., Wang, C., Li, C.L., et al.: The payloads of planetary physics research onboard China's first mars mission (Tianwen-1). Earth Planet. Phys. **4**(4), 331–332 (2020)
2. Ye, P., Sun, Z., Rao, W., Meng, L.: Mission overview and key technologies of the first Mars probe of China. SCIENCE CHINA Technol. Sci. **60**(5), 649–657 (2017). https://doi.org/10.1007/s11431-016-9035-5
3. Li, C.L., Liu, J.J., Geng, Y., et al.: Scientific objectives and payload configuration of China's first Mars exploration mission. J. Deep Space Explor. **5**(5), 406–413 (2018)
4. Ye, B.L., Zhao, J.N., Huang, J.: The status of NASA Mars 2020 rover landing site selection and some thoughts on the landing part of China 2020 Mars mission. J. Deep Space Explor. **4**(4), 310–324 (2017)
5. Ojha, L., Wilhelm, M.B., Murchie, S.L., et al.: Spectral evidence for hydrated salts in recurring slope lineae on Mars. Nat. Geosci. **8**(11), 829–832 (2015)
6. Di, K., Liu, Z., Liu, B., et al.: Topographic analysis of Chang'e-4 landing site using orbital, descent and ground data. In: ISPRS2019, vol. 23, no. 1, pp. 177–180 (2019)
7. Ju, X.W., Wang, Y., Fu, C.L., et al.: Key technology analysis of Mars rover. Space Int. **499**(7), 23–26 (2020)
8. Geng, Y., Zhou, J.X., Li, S., et al.: Review of first Mars exploration mission in China. J. Deep Space Explor. **5**(5), 399–405 (2018)
9. Zhao, J., Wei, S.M., Tang, L., et al.: Review on driving environment of mars rover. Manned Spaceflight **25**(2), 256–267 (2019)
10. Mars exploration rover mission: spotlight (2012). http://marsrover.nasa.gov/
11. Peng, S., Chi, Q., Zheng, Y.: Design of autonomous function scheme for Mars rover. Space Int. **499**(7), 14–18 (2020)
12. Zhou, J.L., Xie, Y., Zhang, Q., et al.: Research on mission planning in teleoperation of lunar rovers. Sci China Info **44**(4), 441–451 (2014)

13. Liu, Z., Di, K., Li, J., et al.: Landing Site topographic mapping and rover localization for Chang'e-4 mission. J. Sci. China-Inf. Sci. **63**, 140901:1-140901:12 (2020)

14. Di, K.C., Jia, M.N., Xin, X., et al.: High-resolution large-area digital orthophoto map generation using LROC NAC images. Photogramm. Eng. Remote. Sens. **85**, 481–491 (2019)

15. Jia, M.N.: High-resolution large-area digital orthophoto map generation based on multi-coverage lunar orbital images and its science application. University of Chinese Academy of Sciences, Beijing (2019)

16. Wang, J., Zhang, Y., Di, K.C., et al.: Localization of the Chang'e-5 lander using radio-tracking and image-based method. Remote Sens. **13**, 590 (2021)

17. Meng, Q.Y., Fu, Z.L., Dong, J.H., et al.: The optical system design of the high-resolution visible spectral camera for China Mars exploration. J. Deep Space Explor. **5**(5), 458–464 (2018)

18. Liu, B., Xu, B., Liu, Z.Q., et al.: Descending and landing trajectory recovery of Chang'e-3 lander using descent images. J. Remote Sens. **18**, 981–987 (2014)

19. Cao, Z., Zhang, Y.J., Ye, W., et al.: Optical design of hazard avoidance camera for Mars rover. Laser Infread **50**(1), 96–100 (2020)

20. Di, K.C., Li, R.: Topographic mapping capability analysis of mars exploration rover 2003 mission imagery. In: Proceedings of the 5th International Symposium on Mobile Mapping Technology (MMT 2007), 28–31 May 2007, Padua, Italy (2007)

21. Di, K.C., Peng, M.: Wide baseline mapping for mars rovers. Photogramm. Eng. Remote. Sens. **77**(6), 609–618 (2011)

22. Peng, M.: High precision and automatic mapping using rover images in deep space exploration. University of Chinese Academy of Sciences, Beijing (2011)

23. Chen, J., Xing, Y., et al.: Guidance, navigation and control technologies of chang'E-3 Lunar Rover. Sci. China Technol. Sci. **44**, 461–469 (2014)

24. Jia, Y., Zhou, Y., Xue, C., et al.: Scientific objectives and payloads of Chang'E-4 mission (in Chinese). Chin. J. Space Sci. **38**(1), 118–130 (2018)

25. Fischler, M.A., Bolles, R.C.: Random sample consensus: a paradigm for model fitting with applications to image analysis and automated cartography. Commun. ACM **24**(6), 381–395 (1981)

26. Liu, B., Niu, S., Xin, X., et al.: High precision DTM and DOM generating using multi-source orbital data on Chang'e-4 landing site. Int. Arch. Photogramm. Remote Sens. Spat. Inf. Sci. 1413–1417 (2019)

27. Robinson, M.S., Brylow, S.M., Tschimmel, M., et al.: Lunar reconnaissance orbiter camera (LROC) instrument overview. Space Sci. Rev. **150**, 81–124 (2010)

Other Research Works and Surveys Related to the Applications of Image and Graphics Technology

A Regularized Limited-Angle CT Reconstruction Model Based on Sparse Multi-level Information Groups of the Images

Lingli Zhang[1,2,3,4](✉), Huichuan Liang[1,2], Xiao Hu[1,2], and Yi Xu[1,2]

[1] Chongqing Key Laboratory of Group and Graph Theories and Applications, Chongqing University of Arts and Sciences, Chongqing, China
Linglizhang@cqwu.edu.cn

[2] Chongqing Key Laboratory of Complex Data Analysis and Artificial Intelligence, Chongqing University of Arts and Sciences, Chongqing, China

[3] Key Lab for OCME, Chongqing Normal University, Chongqing, China

[4] Engineering Research Center of Industrial Computed Tomography Nondestructive Testing of the Education Ministry of China, Chongqing University, Chongqing, China

Abstract. Restricted by the scanning environment and the shape of the target to be detected, the obtained projection data from computed tomography (CT) are usually incomplete, which leads to a seriously ill-posed problem, such as limited-angle CT reconstruction. In this situation, the classical filtered back-projection (FBP) algorithm loses efficacy especially when the scanning angle is seriously limited. By comparison, the simultaneous algebraic reconstruction technique (SART) can deal with the noise better than FBP, but it is also interfered by the limited-angle artifacts. At the same time, the total variation (TV) algorithm has the ability to address the limited-angle artifacts, since it takes into account a priori information about the target to be reconstructed, which alleviates the ill-posedness of the problem. Nonetheless, the current algorithms exist limitations when dealing with the limited-angle CT reconstruction problem. This paper analyses the distribution of the limited-angle artifacts, and it emerges globally. Then, motivated by TV algorithm, tight frame wavelet decomposition and group sparsity, this paper presents a regularization model based on sparse multi-level information groups of the images to address the limited-angle CT reconstruction, and the corresponding algorithm called modified proximal alternating linearized minimization (MPALM) is presented to deal with the proposed model. Numerical implementations demonstrate the effectiveness of the presented algorithms compared with the above classical algorithms.

Keywords: Limited-angle CT · Group sparsity · Wavelet tight frame · Regularization · Proximal alternating linearized minimization

1 Introduction

Computed Tomography (CT) is one of the best nondestructive testing techniques, which shows the internal structure and defect information of the objects to be detected in a

Y. Wang and W. Song (Eds.): IGTA 2021, CCIS 1480, pp. 267–282, 2021.
https://doi.org/10.1007/978-981-16-7189-0_21

nondestructive and accurate way [1]. It is widely used in the field of biomedical industry security inspection and so on. X-ray CT is used to obtain the CT image of the target to be detected, and the target detection is realized through enhanced segmentation and recognition of the CT image. However, due to the different limitations and requirements on CT scanning speed and X-ray dose level in different application scenarios, as well as the limitations of some applications in the field scanning environment and the structure of the target to be detected, the acquired projection data are incomplete and contain a certain level of noise in industrial nondestructive testing (NDT), the scanning of objects (such as castings) by CT system usually does not consider the radiation dose of the detected object. However, due to its particularity, such as the shape of the object being too large or too long, incomplete projection data are often obtained. For example, in the nondestructive testing of pipelines in service [2], the X-ray source and detector can only rotate within a limited-angle scanning range usually less than 180° + fan-angle, which dues to the limitation of the scanning environment such as the surface attached to the pipeline to be tested [3]. In terms of medicine, in view of excessive X-ray radiation to human body may lead to cancer or other diseases [4, 5], CT detection is still one of the important detection methods in medical diagnosis. During the outbreak, for example, many hospitals were preliminary data and reports suggest that a virus nucleic acid testing positive result has a certain lag, experts recommend CT image as the main diagnostic basis currently COVID-19 [6], therefore, for the human body (such as internal organs, organizations, bones, etc.) of the scan, X-ray dose must be strictly controlled. In general, the most direct way to reduce X-ray radiation dose is lower tube voltage or current, but it would introduce additional noise in obtain the projection data [7, 8] and another way is to reduce the number of X ray through the targets to be detected, this will lead to obtain the projection data of scan angle is usually less than 180° + fan-angle. For example, dental CT [9], C-arm CT [10], chest CT and breast CT [11], etc.

In the above scenario, it is difficult to get the complete projection data directly, so from incomplete projection data CT image reconstruction is a ill-posed inverse problem [12, 13] in limited-angle projection data, the reconstructed image from the classical filtered-back projection algorithm (FBP) [1] is affected by limited-angle artifacts, which will lead to the medical diagnosis and non-destructive testing in the late. The basic idea of the regularization method is to approximate the solution of the limited-angle CT reconstruction problem by the solution of a kind of relatively well-posed problem. In a certain extent, optimization method based on regularization not only can alleviate the problem of limited-angle reconstruction which is not qualitative, but also can inhibit limited-angle artifacts and noise compared with classic commercial FBP algorithm.

In recent years, the based regularization optimization reconstruction model has been widely researched and applied [14–17], and the corresponding computing techniques and algorithms have also been developed. Different regularization parameters and different transformations will lead to different models, and the obtained solutions will also be different. Frikel et al. introduced the sparse image under the curvelet sparse transform into the reconstruction process as the prior information, and proposed a regularization method that could stably maintain the boundary, namely the curvelet sparse regularization method CSR [18]. Compared with FBP algorithm and SART technology [12], CSR algorithm improves the quality of limited-angle image reconstruction, but there are still limited-angle artifacts in the reconstructed image. Although it is difficult to completely solve the problem of limited-angle reconstruction, there are methods to analyze the limited-angle artifacts in some feature boundaries of the image [19]. In order to suppress the limited-angle artifacts, many researchers have made some efforts to solve these intractable problems. Take TV algorithm as an example, it has been successfully applied to few-view reconstruction [20]. Although TV algorithms are capable of handling artifacts and perturbations in a small range, they cannot successfully modify global artifacts. due to incomplete limited-angle projection data. Xiaohao et al. improved the first step of the Mumford-Shah (MS) segmentation model [21], and obtained an MS-like model [22], which can obtain a better restored image. Recently, Zeng Li et al. studied an iterative reconstruction algorithm based on wavelet tight frame and ℓ_0 quasi-norm regularization [23], which can better suppress the limited-angle artifacts and protect the edge. However, for the targets with rich features and details, this algorithm can suppress the limited-angle artifacts to a certain extent, but the features and details of the image are blurred to a certain extent.

In order to further research the limited-angle reconstruction problem, this paper presents a regularization model based on sparse multi-level information groups of the images to address the limited-angle CT reconstruction, and the corresponding algorithm called modified proximal alternating linearized minimization (MPALM) is presented to deal with the proposed model. The presented model takes into account several key points: the advantages of TV algorithm, the multi-level of tight frame wavelet decomposition and group sparsity of the transformed image.

The rest of the paper is composed of Sect. 2 to Sect. 6. Section 2 introduces the analysis of group sparsity model. Section 3 gives the proposed model in this paper. Section 4 demonstrates the corresponding algorithms and Sect. 5 exhibits a number of numerical experiments. Finally, conclusions and prospects are made in Sect. 6.

2 Analysis of Group Sparsity Model

In order to deal with the limited-angle CT reconstruction problem, this paper analyses the globally distribution of the limited-angle artifacts and they exist not only in high frequency part but also in low frequency part. Then, there is an urgent need for a transformation that can cover both high and low frequencies, i.e. the multi-level information structure. Therefore, this paper considers the following model:

$$\arg\min_{x} \left\{ \frac{1}{2} \|Ax - b\|_D^2 + \lambda \|Tx\|_{p,q} \right\}. \tag{1}$$

Where x denotes the image to be reconstructed, D is a diagonal matrix, A denotes the system matrix, b is the obtained projection data, T is the sparse transform, $\|\xi\|_D^2 = \xi'D\xi$, 'is the transpose of the matrix, λ is a balancing parameter between the fidelity term and regularization term and $0 \le q \le p \le 2$, and $\|\xi\|_{p,q}$ can be expressed as follows:

$$\|\xi\|_{p,q} = \begin{cases} (\sum_{i=1}^{r} \|\xi_{G_i}\|_p^q)^{1/q} & q > 0, \\ \sum_{i=1}^{r} \|\xi_{G_i}\|_p^0 & q = 0. \end{cases} \tag{2}$$

These are many cases that need to be considered in the proposed model, and here are the key consideration as follows:

1. The choice of the sparse transform T is of great importance. This idea originally came from compressed sensing (CS) [24]. It mainly utilized the prior information of the image to be reconstructed. For the transform T, it can choose a great diversity of the sparse transforms. Especial typicals are gradient transform and wavelet frame decomposition for the image processing (including image restoration, image reconstruction, etc.). Additionally, shearlet transform and curvelet transform are also higher frequency transforms. In this study, wavelet frame decomposition is used to address the limited-angle CT reconstruction problem. It includes a few advantages: firstly, it can be referred to as the generalization of the gradient transform; secondly, it not only includes the high frequency information, but also contains the low frequency information of the reconstructed image (this proposition can deal with the limited-angle artifacts); thirdly, it includes the high order gradients of the multiple directions.

2. The $\ell_{p,q}$ norm includes the group structure. It involves group size and the overlapping nature of the groups. In general, the group size can be selected as $(2l+1) \times (2l+1)$, where l is a natural number. Generally, l can be set to 0, 1 and 2. It's worth noting that the group sparsity not only consider of the transformed image, but also takes into account the structure sparsity. When l is set 1, it degenerates into:

$$\arg\min_x \left\{ \frac{1}{2} \|Ax - b\|_D^2 + \lambda \|Tx\|_q \right\}. \tag{3}$$

Currently, the similar studies have been conducted on above model in references [23]. Suppose that the object Tx is divided into r groups, i.e., $G_i, i = 1, 2, ..., r$, where G_i is the finite set. If $G_i \cap G_j = \Phi$, then they are the overlapping groups.

3. This model needs to consider one important parameter λ. It depends on the image x and the transform T. When T is the gradient transform, the only one parameter λ is properly set to balancing the fidelity term and the regularization term. When T is tight frame decomposition, λ will depend on the level of T. If let L denotes the number of the all levels, then there are L parameters to be identified, i.e., $\lambda_i, i = 1, 2, ..., L$.

4. There's something else to consider is the evaluation of p and q. They are usually set as $0 \le q \le 1 \le p \le 2$. The value of p implies a measure within the elements of

the group. q stands for the measure among the groups. In the existing application studies, p tends to 1 and 2. The other parameter q is generally set to 0 and 1. For the different values of p and q, there will demonstrate different models which can be listed in Table 1. Since the group sparsity is the key idea of the presented model, GSL is utilized to represent the relevant model as shown in Table 1.

Table 1. Types of models.

p	q	Model	Denoted
2	1	$\arg\min\limits_{x}\left\{\frac{1}{2}\|Ax - b\|_D^2 + \lambda\|Tx\|_{2,1}\right\}$	GSL21
2	0	$\arg\min\limits_{x}\left\{\frac{1}{2}\|Ax - b\|_D^2 + \lambda\|Tx\|_{2,0}\right\}$	GSL20
1	1	$\arg\min\limits_{x}\left\{\frac{1}{2}\|Ax - b\|_D^2 + \lambda\|Tx\|_{1,1}\right\}$	GSL11
1	0	$\arg\min\limits_{x}\left\{\frac{1}{2}\|Ax - b\|_D^2 + \lambda\|Tx\|_{1,0}\right\}$	GSL10

One thing to note here is the regularization term, which is the $\ell_{p,q}$ norm of the group structure defined by the model (1). The value of p represents the sparsity among the groups $G_i, i = 1, 2, ..., r$. $p = 0$ is more sparser than $p = 1$.

To solve the limited-angle CT image reconstruction problem, this paper researches the corresponding situations:

$$\{(p, q)|p \in \{1, 2\}, q \in \{0, 1\}\}.$$

The group size considers 1×1, 3×3. The sparse transform is set as wavelet frame decomposition. In all, there are eight situations to be experimented. Corresponding to the limited-angle CT reconstruction, the remaining will not be covered for the time being.

3   The Proposed Model

The proposed model is based on the group sparsity and wavelet frame decomposition, which can be shown as follows:

$$\arg\min_{x}\left\{\frac{1}{2}\|Ax - b\|_D^2 + \sum_{i=1}^{L}\lambda_i\|(Tx)_i\|_{p,q}\right\}. \qquad (4)$$

Where T is selected as wavelet frame decomposition, p is set as 1 or 2, q is set as 0 or 1. The combined treatment of $\ell_{p,q}$ norm and T is more difficult. They need to be decoupled and processed separately. Then inspired by the work of Bin Dong and Jérôme

Bolte etc. [25, 26], by introducing an auxilary variable α converted into the following modality:

$$\underset{x}{\arg\min}\left\{\frac{1}{2}\|Ax - b\|_D^2 + \sum_{i=1}^{L}\lambda_i\|(\alpha)_i\|_{p,q}\right\}, s.t.\alpha = Tx. \tag{5}$$

The solution of the converted form (5) has a lot of ways. The current common solution methods include the Alternating Direction Method of Multipliers (ADMM) [27] and the penalty function method. In this research, the second method is utilized and the model (5) can be converted into the unconstrained form as follows:

$$\underset{x,\alpha}{\arg\min}\left\{\frac{1}{2}\|Ax - b\|_D^2 + \sum_{i=1}^{L}\lambda_i\|(\alpha)_i\|_{p,q} + \frac{\gamma}{2}\|\alpha - Tx\|_2^2\right\}. \tag{6}$$

When $p = 2$ or 1 and $q = 1$, the model (6) is convex and in the convex set, the Gauss-Seidel method [28] is able to address (6) with good convergent results but in order to obtain these good results, the key assumption is that the minimum in every step has a unique solution. When $p = 2$ or 1 and $q = 0$, the model (6) is nonconvex and nonsmooth. So the proximal Alternating Linearized Minimization (PALM) method can be utilized to address (6). But PALM needs minor revision where the number of the parameters has been expanded in the process to be suitable for the proposed model in this paper.

Since the wavelet frame decomposition has L levels. The balancing parameters include $\lambda_i, i = 1, 2, ..., L$, while $\ell_{p,q}$ term of the model (6) has only one parameters λ. In order to facilitate the later numerical experiments, in this research, λ_i can be set as $\lambda * (0.25)^i$. So if λ is confirmed, λ_i is also confirmed.

4 Algorithms

For different models and problems, appropriate algorithms should be adopted to solve them. In order to better solve the model (6), this paper first gives some notations and introduces some related basic definitions and algorithms.

Denote:

$$f(x) = \frac{1}{2}\|Ax - b\|_D^2, \tag{7}$$

$$g(\alpha) = \sum_{i=1}^{L}\lambda_i\|(\alpha)_i\|_{p,q}, \tag{8}$$

$$h(x, \alpha) = \frac{\gamma}{2}\|\alpha - Tx\|_2^2, \tag{9}$$

$$\Psi(x, \alpha) = f(x) + g(\alpha) + h(x, \alpha). \tag{10}$$

Definition 1. [26]: Let $\sigma : R^n \rightarrow (-\infty, +\infty]$ be a proper and lower semi-continuous function. For a given $x \in R^n$ and $u > 0$, the proximal map associated to σ is defined as following:

$$\text{Pr} \, ox_u^\sigma (x) := \underset{y \in R^n}{\arg\min} \left\{ \sigma(y) + \frac{u}{2} \|y - x\|_2^2 \right\}. \tag{11}$$

A. PALM Algorithm

The PALM algorithm can be viewed as alternating the steps of the well-known proximal-forward-backward (PFB) scheme. As is known to all, PFB scheme is to minimize the sum of a smooth function with a nonsmooth function. For dealing with the model (6), the flowchart of PALM algorithm can be given as Algorithm 1 [26].

Algorithm 1: PALM (Proximal Alternating Linearized Mnimization)

Step 1. Initialize: Given any (x^0, α^0);

Step 2. For the generated sequence $\{(x^k, \alpha^k)\}$:

Given $\gamma_1 > 1$, let $c_k = \gamma_1 L_1(\alpha^k)$ and compute:

$$x^{k+1} \in \text{Pr} \, ox_{c_k}^f (x^k - \frac{1}{c_k} \nabla_x h(x^k, \alpha^k))$$

Given $\gamma_2 > 1$, let $d_k = \gamma_2 L_2(x^{k+1})$ and compute:

$$\alpha^{k+1} \in \text{Pr} \, ox_{d_k}^g (\alpha^k - \frac{1}{d_k} \nabla_\alpha h(x^{k+1}, \alpha^k))$$

J·Bolte et al. [26] utilized the Kurdyka-Lojasiewicz (KL) property to derive a critical point of the bounded sequence generated by PALM algorithm. This conclusion also needs some corresponding assumptions, like assumptions in the reference [26].

B. Assumptions

Assumption 1: $f : R^n \rightarrow (-\infty, +\infty]$ are proper and lower semi-continuous functions. $h : R^n \times R^m \rightarrow R$ is a continuously differentiable function, where m depends on the sparse transform T.

Assumption 2: (i) $\underset{R^n \times R^m}{\inf} \Psi > -\infty$, $\underset{R^n}{\inf} f > -\infty$ and $\underset{R^m}{\inf} g > -\infty$.

(ii) For any given $\alpha, x_1, x_2 \in R^n$, the function $h(x, \alpha)$ is $C_{L_1(\alpha)}^{1,1}$, namely the partial gradient $\nabla_x h(x, \alpha)$ is globally Lipschitz with moduli $L_1(\alpha)$ as following:

$$\|\nabla_x h(x_1, \alpha) - \nabla_x h(x_2, \alpha)\|_2 \leq L_1(\alpha) \|x_1 - x_2\|_2$$

For any given $x, \alpha_1, \alpha_2 \in R^m$, the function $h(x, \alpha)$ is $C_{L_2(x)}^{1,1}$, namely the partial gradient $\nabla_\alpha h(x, \alpha)$ is globally Lipschitz with moduli $L_2(x)$ as following:

$$\|\nabla_\alpha h(x, \alpha_1) - \nabla_\alpha h(x, \alpha_1)\|_2 \leq L_2(x) \|\alpha_1 - \alpha_2\|_2$$

(iii) According to the sequences $\{L_1(\alpha^k)\}$ and $\{L_2(x^k)\}$, there exist $\rho_i^-, \rho_i^+ > 0, i = 1, 2$ such that:

$$\inf\{L_1(\alpha^k) : k \in N\} \geq \rho_1^-, \sup\{L_1(\alpha^k) : k \in N\} \leq \rho_1^+,$$

$$\inf\{L_2(x^k) : k \in N\} \geq \rho_2^-, \sup\{L_2(x^k) : k \in N\} \leq \rho_2^+.$$

C. MPALM (Modified Proximal Alternating Linearized Minimization)

Based on the PALM scheme, it is theoretically effective to the limited-angle CT image reconstruction problem [29], the inverse matrix $(A'DA + c_kI)^{-1}$ is difficult to obtain, so it needs to modify the PALM scheme for the current troublesome problem. Then the definition of the modified proximal mapping can be given as following:

Definition 2 [30]: Let $\sigma : R^n \to (-\infty, +\infty]$ be a proper and lower semi-continuous function. Let $V \in R^{n \times n}$ be a symmetric positive definite matrix. For any given $x \in R^n$, the proximal map associated to σ with the induced V can be defined as:

$$\Pr ox_V^\sigma(x) := \arg\min_{y \in R^n}\left\{\sigma(y) + \frac{1}{2}\|y - x\|_V^2\right\}. \tag{12}$$

Including above definition, the modified PALM (MPALM) can be given as following:

Algorithm 2: MPALM (Modified Proximal Alternating Linearized Minimization)

Step 1. Initialize: Given any (x^0, α^0);

Step 2. For the generated sequence $\{(x^k, \alpha^k)\}$:

Given $\omega > 1$, let $c_k = \omega L_1$ and compute:

$$x^{k+1} \in \Pr ox_V^{h(x, \alpha^k)}(x^k - V^{-1}\nabla_x f(x^k))$$

Given $d > 0$ and compute:

$$\alpha^{k+1} \in \Pr ox_d^{g(\alpha)}(Tx^{k+1})$$

By modifying PALM, it has the ability of dealing with the limited-angle CT image reconstruction problem. It can also avoid computing $(A'DA + c_kI)^{-1}$. Then it can obtain the exact minimization of the first problem of Algorithm 2 as shown in:

$$x^{k+1} = (\gamma I + V)^{-1}[V(x^k - V^{-1}A'D(Ax^k - b)) + \gamma T'\alpha^k]. \tag{13}$$

Where $\gamma I + V$ and V are diagonal matrix. The minimization of the second problem of Algorithm 2 depends on p and q.

(1) When $p = 2$ and $q = 1$, the soft thresholding (ST) [31] can deal with that, for $i = 1, 2, ..., L, \beta_{G_i} = (Tx^{k+1})_{G_i}$, the solution is given as following:

$$\alpha_{G_i}^{k+1} = \begin{cases} (1 - \dfrac{\lambda_i/d}{\|\beta_{G_i}\|_2})\beta_{G_i}, & \|\beta_{G_i}\|_2 > \lambda_i/d, \\ 0, & otherwise. \end{cases} \tag{14}$$

(2) When p = 2 and q = 0, the hard thresholding (HT) [32] can deal with that, for $i = 1, 2, ..., L$, $\beta_{G_i} = (Tx^{k+1})_{G_i}$, the solution is given as following:

$$
\alpha_{G_i}^{k+1} = \begin{cases} \beta_{G_i}, & \|\beta_{G_i}\|_2 > \sqrt{2\lambda_i/d}, \\ \{0, \beta_{G_i}\}, & \|\beta_{G_i}\|_2 = \sqrt{2\lambda_i/d}, \\ 0, & \textit{otherwise}. \end{cases} \tag{15}
$$

(3) When p = 1 and q = 1, ST can deal with that, for $i = 1, 2, ..., L$, $\beta_{G_i} = (Tx^{k+1})_{G_i}$, the solution is given as following:

$$
\alpha_{G_i}^{k+1} = \beta_{G_i} - \frac{2\lambda_i}{d} sign(\beta_{G_i}). \tag{16}
$$

(4) When p = 1 and q = 0, HT can deal with that, for $i = 1, 2, ..., L$, $\beta_{G_i} = (Tx^{k+1})_{G_i}$, the solution is given as following:

$$
\alpha_{G_i}^{k+1} = \begin{cases} \beta_{G_i}, & \|\beta_{G_i}\|_1 > \sqrt{2\lambda_i/d}, \\ \{0, \beta_{G_i}\}, & \|\beta_{G_i}\|_1 = \sqrt{2\lambda_i/d}, \\ 0, & \textit{otherwise}. \end{cases} \tag{17}
$$

5 Numerical Experiments

In order to check the effectiveness of the presented model and responding algorithm, a digital NURBS based cardiac torso (NCAT) phantom [33] is utilized which can be shown as Fig. 1. To analyze the influence of the limited-angle artifacts, the experiments had not taken the other factors into account. The projection data can be generated by projecting a 256×256 discretized NCAT phantom from the scanning angles [0, 160°], [0, 140°], [0, 120°] and [0, 100°]. The corresponding simulated geometrical scanning parameters can be shown in Table 2.

For the numerical experiments, this paper utilizes the presented algorithms (include GSL21, GSL20, GSL11 and GSL10) to implement the simulated experiments, compared with FBP [12], SART [12], TV [20]. To reveal the details and edges in the reconstructed results, the local areas have been magnified. In addition, the quantitative assessments have been utilized to verify the effectiveness of the above algorithms by the root mean squared error (RMSE) [34], peak signal to noise ratio (PSNR) [34] and structural similarity index (SSIM) [35] as shown in formulas (18)–(20).

$$
RMSE = \sqrt{\frac{1}{M \times N} \sum_{i=1}^{N} \sum_{j=1}^{M} \left(x_{i,j} - x_{i,j}^r\right)^2} \tag{18}
$$

$$
PSNR = 10\log_{10} \frac{\|x^r\|_\infty^2}{RMSE^2} \tag{19}
$$

$$
SSIM(x, x^r) = \frac{2\mu_x\mu_{x^r}(2Cov\{x, x^r\} + c_2)}{(\mu_x^2 + \mu_{x^r}^2 + c_1)(\sigma_x^2 + \sigma_{x^r}^2 + c_2)} \tag{20}
$$

There into, x^r is the reference image; x shows the reconstructed result; $M \times N$ denotes the total number of the image pixels; μ_x and μ_{x^r} are the mean values of x and x^r, respectively; σ_x and σ_{x^r} are the standard deviations of x and x^r; $Cov\{x, x^r\}$ denotes the covariance between x and x^r; $c_1 = (0.01R)^2$ and $c_2 = (0.03R)^2$, R is the dynamic range of pixel values. The closer to 1 the value of SSIM is, the higher similarity between the reference image and the reconstructed image is.

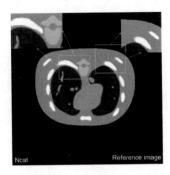

Fig. 1. It is the reference image and the rectangles stand for the locally zoom-in positions.

Table 2. The scanning parameters of simulated CT imaging system for NCAT

Parameter	Value
The distance from X-ray source to rotation center	900.0 mm
The distance from X-ray source to detector	1300.0 mm
Interval angle between two adjacent projection views	0.6679°
The number of detector units	256
Image size	256 × 256
Pixel size of the image	1.0 × 1.0mm2

In numerical experiments, the parameters are chose by trial and error for better image quality, and the relevant values will be in quantitative table. $N_{\max}$ stands for the number of iterations. N_{TV} is the number of TV minimization step. ω denotes the relaxation parameter of SART. λ, γ and d are the parameters in model (6). The model adopts the non-overlapping group 1×1 and 3×3.

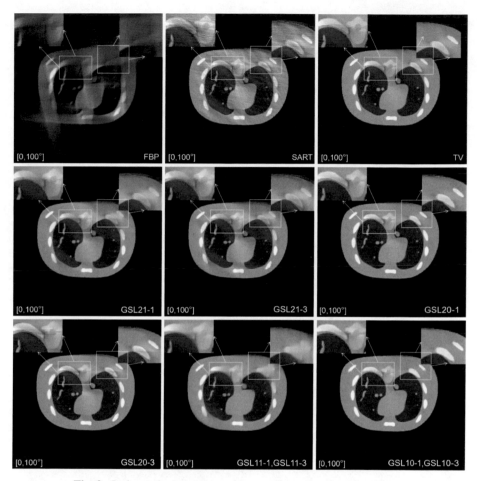

Fig. 2. It shows the reconstructed images from the given algorithms.

A total of four numerical experiments were conducted. One of these experiments is shown here like Fig. 2. It demonstrates the images reconstructed using FBP, SART, TV, GSL21, GSL20, GSL11 and GSL10 algorithms from the scanning scope [0,100°], and the corresponding algorithm is indicated in the lower right corner of the Fig. 2, where GSL21-1 denotes the GSL21 algorithm adopts the 1×1 group and the others are similar. It can be seen that the results by TV, GSL20 and GSL10 has the ability of addressing the limited-angle artifacts while preserving the details and edges, which also indicates the advantages of the ℓ_0 regularization. The results reconstructed using FBP and SART suffer from the artifacts and noise seriously. In order to better reflect the difference of the above results, the histogram and quantitative assessments are utilized as shown in Fig. 3 and Table 3. Figure 3 represents the 83th row histogram of the reconstructed images using nine algorithms from the scanning angle [0,100°]. The histogram image indicates that the GSL10 is capable of dealing with the limited-angle artifacts and it is the closest

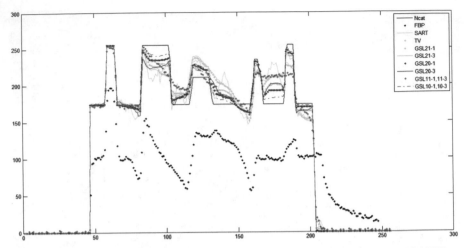

Fig. 3. It shows the 83th row histogram of the images from the scanning angle [0,100°].

the reference image. The quantitative assessments in Table 3 also demonstrates that the result by GSL10 can obtain the best evaluation.

Table 3. Quantitative assessments of the reconstructed images and the corresponding parameters.

Angle	Algorithm	Parameter	RMSE	PSNR	SSIM	UQI
[0, 160°]	FBP	None	21.6661	21.0233	0.6684	0.9428
	SART	Nmax = 200, ω = 1.0	3.4509	37.3723	0.9854	0.9989
	TV	Nmax = 200, ω = 1.0, Ntv = 20, λ = 0.2	2.0235	42.0087	0.9968	0.9996
	GSL21-1	Nmax = 200, ω = 1.0, λ = 0.0005, γ = 0.8, d = 0.002	2.5342	40.0541	0.9953	0.9994
	GSL21-3	Nmax = 200, ω = 1.0, λ = 0.001, γ = 0.8, d = 0.002	2.9970	38.5970	0.9936	0.9992
	GSL20-1	Nmax = 400, ω = 1.0, λ = 0.008, γ = 0.8, d = 0.002	1.7479	43.2805	0.9967	0.9997
	GSL20-3	Nmax = 600, ω = 1.0, λ = 0.003, γ = 0.8, d = 0.002	2.0627	41.8419	0.9958	0.9996
	GSL11-1	Nmax = 600, ω = 1.0, λ = 0.001, γ = 0.8, d = 0.002	2.8101	39.1564	0.8899	0.9993
	GSL10-1	Nmax = 1200, ω = 1.0, λ = 0.001, γ = 0.8, d = 0.002	**0.4296**	**55.4688**	**0.9995**	**1.0000**
[0, 140°]	FBP	None	29.5853	18.7093	0.6466	0.8984

(continued)

Table 3. (*continued*)

Angle	Algorithm	Parameter	RMSE	PSNR	SSIM	UQI
	SART	Nmax = 300, ω = 1.0	5.3814	33.5128	0.9755	0.9973
	TV	Nmax = 300, ω = 1.0, Ntv = 20, λ = 0.2	3.2616	37.8622	0.9942	0.9990
	GSL21-1	Nmax = 300, ω = 1.0, λ = 0.0005, γ = 0.8, d = 0.002	4.3798	35.3018	0.9912	0.9982
	GSL21-3	Nmax = 300, ω = 1.0, λ = 0.001, γ = 0.8, d = 0.002	4.9463	34.2453	0.9889	0.9977
	GSL20-1	Nmax = 500, ω = 1.0, λ = 0.008, γ = 0.8, d = 0.002	2.3210	40.8175	0.9943	0.9995
	GSL20-3	Nmax = 800, ω = 1.0, λ = 0.003, γ = 0.8, d = 0.002	3.8192	36.4914	0.9920	0.9986
	GSL11-1	Nmax = 800, ω = 1.0, λ = 0.001, γ = 0.8, d = 0.002	3.6001	37.0045	0.8860	0.9988
	GSL10-1	Nmax = 1400, ω = 1.0, λ = 0.001, γ = 0.8, d = 0.002	**1.1913**	**46.6101**	**0.9982**	**0.9999**
[0, 120°]	FBP	None	35.5614	17.1112	0.6476	0.8485
	SART	Nmax = 400, ω = 1.0	7.7719	30.3203	0.9557	0.9943
	TV	Nmax = 400, ω = 1.0, Ntv = 20, λ = 0.2	4.7504	34.5962	0.9919	0.9979
	GSL21-1	Nmax = 400, ω = 1.0, λ = 0.0005, γ = 0.8, d = 0.002	6.9805	31.2530	0.9851	0.9954
	GSL21-3	Nmax = 400, ω = 1.0, λ = 0.001, γ = 0.8, d = 0.002	7.6527	30.4545	0.9823	0.9945
	GSL20-1	Nmax = 800, ω = 1.0, λ = 0.001, γ = 0.8, d = 0.002	4.6196	34.8387	0.9917	0.9980
	GSL20-3	Nmax = 1000, ω = 1.0, λ = 0.003, γ = 0.8, d = 0.002	5.3449	33.5720	0.9887	0.9973
	GSL11-1	Nmax = 1000, ω = 1.0, λ = 0.001, γ = 0.8, d = 0.002	5.8517	32.7852	0.8787	0.9968
	GSL10-1	Nmax = 1600, ω = 1.0, λ = 0.001, γ = 0.8, d = 0.002	**1.8614**	**42.7341**	**0.9970**	**0.9997**
[0, 100°]	FBP	None	42.4681	15.5695	0.6521	0.7624
	SART	Nmax = 500, ω = 1.0	10.5055	27.7025	0.9137	0.9896
	TV	Nmax = 500, ω = 1.0, Ntv = 20, λ = 0.2	5.7517	32.9350	0.9890	0.9969
	GSL21-1	Nmax = 500, ω = 1.0, λ = 0.0005, γ = 0.8, d = 0.002	9.3168	28.7455	0.9762	0.9918
	GSL21-3	Nmax = 500, ω = 1.0, λ = 0.001, γ = 0.8, d = 0.002	10.0004	28.1305	0.9732	0.9905

(*continued*)

Table 3. (*continued*)

Angle	Algorithm	Parameter	RMSE	PSNR	SSIM	UQI
	GSL20-1	Nmax = 1400, ω = 1.0, λ = 0.001, γ = 0.8, d = 0.002	4.8702	34.3799	0.9912	0.9978
	GSL20-3	Nmax = 1400, ω = 1.0, λ = 0.003, γ = 0.8, d = 0.002	6.3227	32.1128	0.9870	0.9962
	GSL11-1	Nmax = 1400, ω = 1.0, λ = 0.001, γ = 0.8, d = 0.002	7.7149	30.3842	0.8695	0.9944
	GSL10-1	Nmax = 1800, ω = 1.0, λ = 0.001, γ = 0.8, d = 0.002	**3.6386**	**36.9121**	**0.9941**	**0.9988**

6   Conclusions and Prospects

In order to address the limited-angle CT reconstruction problem, this paper investigates a regularization model based on sparse multi-level information groups of the images, which takes TV algorithm, tight frame wavelet decomposition and group sparsity into consideration, and the modified proximal alternating linearized minimization (MPALM) is presented to deal with the proposed model. Although, to some extent, it has the ability to deal with the limited-angle artifacts, the corresponding parameters are chose by trial and error and the size of group sparsity should be researched in the future, and a random group will be used in the presented model.

Acknowledgments. This work is supported by the Science and Technology Research Program of Chongqing Municipal Education Commission (Grant No. KJQN2019013), the Scientific Research Foundation of Chongqing University of Arts and Sciences (Grant No. R2019FSC17), the Natural Science Foundation of Chongqing Municipal Science and Technology Commission (Grant numbers: cstc2020jcyj-msxm2352), and the Open Project of Key Laboratory No.CSSXKFKTQ202004, School of Mathematical Sciences, Chongqing Normal University.

References

1. Era, T.: Computed Tomography (2021)
2. Yumeng, G., Li, Z., et al.: Image reconstruction method for exterior circular cone-beam CT based on weighted directional total variation in cylindrical coordinates. J. Inverse Ill-posed Prob. **28**(2), 155–172 (2020)
3. Noo, F., et al.: Image reconstruction from fan-beam projections on less than a short scan. Phys. Med. Biol. **47**(14), 2525–2546 (2002)
4. Brenner, D.J., Hall, E.J.: Computed tomography–an increasing source of radiation exposure. N. Engl. J. Med. **357**(22), 2277–2284 (2013)
5. Berrington, G.D.A., et al.: Projected cancer risks from computed tomographic scans performed in the United States in 2007. Arch. Intern. Med. **169**(22), 2071–7 (2009)
6. Jiansong, J., Tiemin, W., Weibin, Y., Zufei, W.: Early CT Signs and Differential Diagnosis of COVID-19. Science Press (2020)

7. Hsieh, J.: Adaptive streak artifact reduction in computed tomography resulting from excessive x-ray photon noise. Med. Phys. **25**(11), 2139–2147 (1998)
8. Li, T., et al.: Nonlinear sinogram smoothing for low-dose X-ray CT. IEEE Trans. Nucl. Sci. **51**(5), 2505–2513 (2004)
9. D'Souza, K., Aras, M.: Applications of CAD/CAM technology in dental implant planning and implant surgery. In: Chaughule, R.S., Dashaputra, R. (eds.) Advances in Dental Implantology using Nanomaterials and Allied Technology Applications, pp. 247–286. Springer, Cham (2021). https://doi.org/10.1007/978-3-030-52207-0_11
10. Yinsheng, L., Garrett, J.W., Ke, L., Charles, S., Guang-Hong, C.: An enhanced smart-recon algorithm for time-resolved c-arm cone-beam ct imaging. IEEE Trans. Med. Imaging **39**(6), 1894–1905 (2020)
11. He, Q., Ma, L., You, C.: Letter by He et al regarding article, "signs of pulmonary infection on admission chest computed tomography are associated with pneumonia or death in patients with acute stroke". Stroke, **51**(11) (2020)
12. Buzug, T.M.: Computed Tomography: from Photon Statistics to Modern Cone-Beam CT. Springer, Leipzig (2008). https://doi.org/10.1007/978-3-540-39408-2
13. Natterer, F.: The Mathematics of Computerized Tomography. Wiley, New York (1986)
14. Sidky, E.Y., Pan, X.: Image reconstruction in circular cone-beam computed tomography by constrained, total-variation minimization. Phys. Med. Biol. **53**, 4777–4807 (2008)
15. Frikel, J.: Sparse regularization in limited angle tomography. Appl. Comput. Harmon. Anal. **34**(1), 117–141 (2013)
16. Frikel, J.: A new framework for sparse regularization in limited angle x-ray tomography. In: IEEE International Symposium on Biomedical Imaging: From Nano To Macro, pp. 824–827. IEEE Press, Piscataway (2010)
17. Li, Z., Jiqiang, G., Baodong, L.: Limited-angle cone-beam computed tomography image reconstruction by total variation minimization and piecewise-constant modifcation. J. Inverse Ill-Posed Probl. **21**(6), 735–754 (2013)
18. Frikel, J., Quinto, E.T.: Characterization and reduction of artifacts in limited angle tomography. Inverse Prob. **29**(12) 125007 (2013)
19. Quinto, E.T.: Artifacts and visible singularities in limited data x-ray tomography. Sens. Imaging **18**(1), 9 (2017)
20. Sidky, E.Y., Kao, C., Pan, X.: Accurate image reconstruction from few-views and limited-angle data in divergent-beam CT. J. Xray Sci. Technol. **14**, 119–139 (2006)
21. Mumford, D., Shah, J.: Optimial approximation by piecewise smooth functions and associated variational problems. Comm. Pure Appl. Math. **42**, 577–685 (1989)
22. Cai, X., Chan, R., Zeng, T.: A two-stage image segmentation method using a convex variant of the Mumford-Shah model and thresholding. SIAM J. Imag. Sci. **6**(1), 368–390 (2013)
23. Li, Z., Wang, C.: Error bounds and stability in the l0 regularized for CT reconstruction from small projections. Inverse Prob. Imaging **10**(3), 829–853 (2017)
24. Boyer, C., et al.: Compressed sensing with structured sparsity and structured acquisition. Appl. Comput. Harmon. Anal. **46**(2), 312–350 (2019)
25. Bin, D., Yong, Z.: An efficient algorithm for l0 minimization in wavelet frame based image restoration. J. Sci. Comput. **54**(2–3), 350–368 (2013). https://doi.org/10.1007/s10915-012-9597-4
26. Bolte, J., Sabach, S., Teboulle, M.: Proximal alternating linearized minimization for nonconvex and nonsmooth problems. Math. Program. **146**(1–2), 459–494 (2013). https://doi.org/10.1007/s10107-013-0701-9
27. Boyd, S.: Alternating direction method of multipliers (2011)
28. Gauss-Seidel Method (2010)

29. Wang, C., Luo, X., Wei, Y., Guo, Y., Zhang, L.: A variational proximal alternating linearized minimization in a given metric for limited-angle CT image reconstruction. Appl. Math. Modell. **67**, 315–336 (2019)
30. Frankel, P., Garrigos, G., Peypouquet, J.: Splitting methods with variable metric for KL functions. J. Optim. Theory Appl. **165**(3), 874–900 (2015)
31. Bredies, K.: Soft-thresholding. J. Fourier Anal. Appl. **14**(5–6), 813–837 (2008)
32. Blumensath, T., Yaghoobi, M., Davies, M.E.: Iterative hard thresholding and L0 regularisation. In: IEEE International Conference on Acoustics. IEEE (2007)
33. Segars, W.P.: Development and application of the new dynamic NURBS-based cardiac-torso (NCAT) phantom. The University of North Carolina at Chapel Hill (2001)
34. Storath, M., Weinmann, A., Frikel, J., et al.: Joint image reconstruction and segmentation using the Potts model. Inverse Prob. **31**(2), 25003–25031(29) (2014)
35. Chow, L.S., Paramesran, R.: Review of medical image quality assessment. Biomed. Signal Process. Control **27**, 145–154 (2016)

Author Index

Printed in the United States
by Baker & Taylor Publisher Services